(A. 10779)

NATURAL ENVIRONMENT RESEARCH COUNCIL
INSTITUTE OF GEOLOGICAL SCIENCES

British Regional Geology

South Wales

(THIRD EDITION)

By T. Neville George, D.Sc., F.R.S.
(Professor of Geology in the University of Glasgow)

Based on previous editions by
J. Pringle, D.Sc.
and T. Neville George

LONDON   HER MAJESTY'S STATIONERY OFFICE   1970

The Institute of Geological Sciences
was formed by the
incorporation of the Geological Survey of Great Britain
and the Museum of Practical Geology
with Overseas Geological Surveys
and is a constituent body of the
Natural Environment Research Council

© *Crown copyright 1970*

First published 1937
Third edition 1970

SBN 11 880084 1

Foreword to Third Edition

Since the publication of the second edition of the 'South Wales' handbook in 1948 much progress has been made in elucidating the stratigraphy and structure of the Carboniferous rocks, especially in and around the main coalfield. There have also been some advances in knowledge of the stratigraphy and palaeogeography of the Ordovician and Silurian rocks, the Old Red Sandstone, and the Lias, and in geomorphological interpretation of the landscape. The Pre-Cambrian and Cambrian rocks have remained neglected.

Although the plan followed in earlier editions has been retained as far as possible in this edition the author has rewritten most of the text and has much improved and amplified the illustrations. The text has been edited by Dr. J. R. Earp, the palaeontology by Mr. R. V. Melville and his colleagues.

The permission of the Trustees of the British Museum (Natural History) to reproduce line blocks shown here as Fig. 27c and J, from *British Palaeozoic Fossils* and Fig. 36B and F, from *British Mesozoic Fossils* is gratefully acknowledged. Fig. 27B is from J. Weir and D. Leitch, 1936: 'The zonal distribution of the non-marine lamellibranchs in the Coal Measures of Scotland', *Trans. Roy. Soc. Edin.*, **58**, 697–751.

An EXHIBIT illustrating the geology and scenery of the region described in this volume is set out in the Museum of Practical Geology, Institute of Geological Sciences, Exhibition Road, South Kensington, London S.W.7.

Contents

Contents

Illustrations

Figures in Text

Plates

---

[1]Numbers preceded by A refer to photographs in the Geological Survey collection.

Plate *xii* **Facing page**

1. Introduction

South Wales is not readily defined topographically except by the sea of Cardigan Bay and the Bristol Channel. Northwards the Plynlimon range merges unbrokenly into the mountains of North Wales. On the east, the Breconshire plain and the Brecon Beacons are structurally and physically one with the Herefordshire plain, the hills of Radnor Forest and Clun Forest, and the Black Mountains. Only in the south-east, where the bold marginal escarpment of the South Wales coalfield is strongly developed, is there a prominent geological line forming a natural boundary.

The region, in arbitrary definition, has a convenient northern boundary at the foot of the Plynlimon range, where the valleys of the lower Dyfi and the Carno may be considered to divide the northern from the southern mountains. An eastern boundary, following the scarp of the South Wales coalfield from near Newport to Abergavenny, continues northwards across a spur of the Black Mountains to the Wye Valley at Talgarth, follows the River Wye towards Builth, and then runs northwards to Newtown. (*See* Fig. 1.)

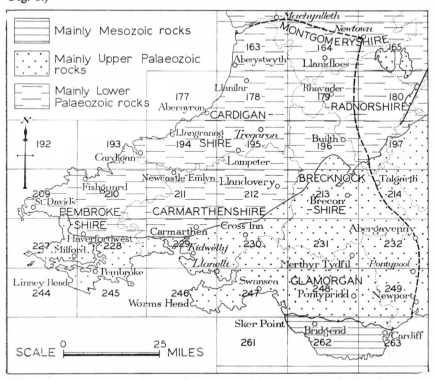

Fig. 1. *Outline map showing the three principal geological regions of South Wales, with a key to the one-inch maps of the Geological Survey*

Geological History

South Wales as a whole is a region mainly of Palaeozoic rocks. All systems (Permian excepted) from Pre-Cambrian to Mesozoic are represented, but Pre-Cambrian rocks crop out only in small areas in Pembrokeshire, and Mesozoic rocks are limited to thin uppermost Trias and lowest Jurassic. Rocks of Cretaceous and, with a minute exception, Tertiary age are lacking, the next succeeding deposits being the superficial glacial drifts of Pleistocene times (*see* Fig. 2).

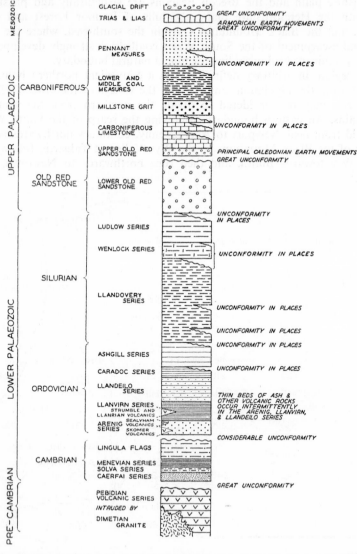

Fig. 2. *Stratigraphical column of the rock sequence in South Wales, simplified and generalized*
(Approximate scale: 1 in to 10 000 ft)

The Pre-Cambrian rocks are almost wholly of igneous origin. The Lower Palaeozoic formations in contrast are mainly marine sediments with inter-bedded volcanic lavas and tuffs. They were deposited in a major basin or geosyncline, formed by subsidence of the floor of Pre-Cambrian rocks, that extended northwards into North Wales and eastwards into England, and that persisted from the beginning of Cambrian until the close of Silurian times. Subsidence was pulsatory, however, downward movement alternating with uplift, and the sea-floor repeatedly emerged and was locally subjected to intense erosion.

The transformation of the Silurian seas into the 'continental' deltas in which the Devonian Old Red Sandstone was deposited was a product of locally powerful earth-movements, of the Caledonian orogeny, that caused a southward retreat of the sea from the Welsh area by the elevation of a land-mass, 'St. George's Land', composed of folded Lower Palaeozoic rocks. The land-mass formed much of central and south-west Wales and perhaps extended to the mountains of Leinster in Ireland. On its flanks the Old Red Sandstone was laid down as river and floodplain deposits that rest over much of their outcrop with marked discordance on the folded and eroded Lower Palaeozoic rocks beneath. A major pulse of earth-movement, a concluding phase of the Caledonian orogeny, occurred during mid-Devonian times when there was revived uplift of St. George's Land, Middle Old Red Sand-stone not being known in Wales. Upper Old Red Sandstone thus every-where rests with unconformity on the rocks beneath, as a sign of late-Devonian subsidence and renewed deposition on the southern flanks of St. George's Land. The earlier members of the Upper Old Red Sandstone are wholly of freshwater origin, but some of the later members contain marine fossils and indicate an encroachment of the sea from the south—the beginnings of the marine advance that is marked in major expression by the limestones of the overlying Carboniferous rocks.

The Lower Carboniferous rocks (the Carboniferous Limestone) are relatively pure shallow-water marine sediments, many of them richly fossiliferous, that are in strong contrast to the Old Red Sandstone beneath. At the time of their formation the seas flanking St. George's Land were relatively free of coarse detritus, but in Upper Carboniferous times limestones diminished in importance in the rock sequence, and were replaced by grits, sandstones, and shales. These terrigenous rocks continued in great part to be deposited in a marine environment during the accumulation of the Millstone Grit; but the Coal Measures, the latest Carboniferous rocks, are almost without marine fossils and show every sign of being the deposits of freshwater or brackish-water deltas and swamps in which peats formed from the debris of lush forest growths were recurrent, in due course to become lithified as coal-seams. (*See* Fig. 3.)

Before the close of Palaeozoic times the region was again affected by powerful earth-movements of the Hercynian orogeny, and was uplifted and intensely folded to form corrugated mountain ranges running from southern Ireland through South Wales into England and forming part of the east-and-west Armorican arc of north-western Europe. When the movements died down, a cycle of erosion, submergence, and sedimentation was initiated, an aspect of which was the incoming of plants and animals characteristic of

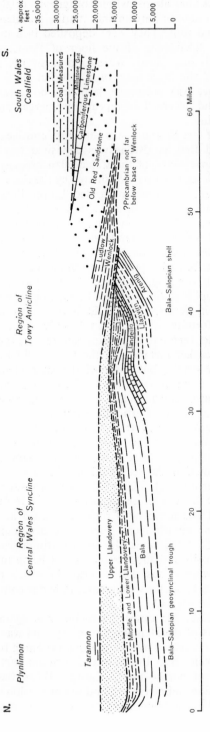

FIG. 3.   *Diagrammatic representation of the sedimentary history of the Palaeozoic rocks in South Wales*

The effects of structural deformation are neglected in order to emphasize the contrast between the thick accumulations of Lower Palaeozoic rocks in mid-Wales, where their total thickness may reach 38 000 ft, but where Upper Palaeozoic rocks are absent, and the thick accumulations of Upper Palaeozoic rocks in South Wales, where they may reach a total thickness of 24 000 ft, but where Lower Palaeozoic rocks diminish to a few thousand feet and may locally be absent. The contrast is reflected in the unconformable contact of the Upper with the Lower Palaeozoic rocks.

the Mesozoic era. Against the southern flanks of the deeply eroded and pared-down Hercynian mountains the Triassic and Liassic sediments were deposited in a basin that extended from southern Pembrokeshire into Gower and south Glamorgan. The red Trias consists largely of material derived in a semi-arid environment from the subaerial waste of the Old Red Sandstone and Carboniferous rocks. As the land sank the river and lake muds were succeeded by grey and blue clays and limestones laid down in the open waters of the advancing Liassic sea.

Little is known of the geological history of the region from Lias times until the onset of glaciation in the Pleistocene period. No trace of post-Liassic Mesozoic rocks has been found, but younger Jurassic rocks may well have been deposited on the Lias, and it is possible that before the close of the era much of the region was drowned beneath the Cretaceous sea, and was covered by the Chalk. The absence of post-Liassic deposits is a measure of the intensity of the denudation that has taken place since early Jurassic times.

Structure

As a result of repeated regional compression all the solid rocks in South Wales are folded into anticlines and synclines and broken by faults and thrusts. The oldest rocks display the greatest deformation, both in the sharpness of folding and the magnitude of fracture, and in the imposition of cleavage on some of the shaly rocks; but even in the Lias the folds have amplitudes of several hundred feet and locally show signs of shearing.

In broad distribution of outcrops, reflections of original stratigraphical relationships have not been altogether obliterated by later structural changes (*see* Pl. XIII). St. George's Land is still to be discerned in the wide expanse of Lower Palaeozoic rocks towards the north and west, where the folds of the Caledonian orogeny (*see* Pl. IXA) are clearly delineated by the inliers of Pre-Cambrian and Ordovician igneous rocks in Pembrokeshire and by the tongues of Ordovician rocks that, flanked by Silurian, run north-eastwards far into mid-Wales as the plunging Teifi and Towy anticlines. To the south, the east-and-west Armorican arc is picked out by the complex synclines of the Pembrokeshire and South Wales coalfields, and by the corrugations of south Pembrokeshire and south Glamorgan in which multiple folds disclose Old Red Sandstone (sometimes Lower Palaeozoic rocks) along their crests, Upper Carboniferous rocks in their troughs (*see* Pl. V). Yet farther south, the coastal limits of the Triassic lake and the Liassic sea are locally preserved in Mesozoic cliffs eroded in Carboniferous rocks, sometimes in almost perfect detail (*see* Pl. XA).

Physical Features and Scenery

In many ways, though not completely, the topography and scenery of South Wales reflect the geological structure and the geological history of the rocks.

In the Lower Palaeozoic terrain north-west of the Towy anticline, grits and hard shales form high bleak moorland with only a thin capping of poor soil. Except in the valleys there is little agriculture, and the population

of sheep farmers is sparse. The relatively isolated region of Mynydd Preseli
in Pembrokeshire presents much the same appearance. Although there are
acute folds in this tract of older rocks, they are not generally marked by
obvious topographical features, and the grain of the country, partly because
of the uniform character of the Ordovician and Silurian sediments, is not
well brought out by the distribution of hill and valley. The principal sign
of adjustment to structure is the Towy valley, which for much of its length
closely follows the axial line of the Towy anticline and which as a strike
depression owes its existence directly to a fluvial etching of the Caledonian
frame. The neighbouring Cothi valley similarly follows the axis of the
Cothi anticline. On the other hand, the subparallel Teifi valley is not so
closely adjusted to the geology, for its course lies a mile or two south-east
of the main arch of the Teifi anticline, and between Llandyssul and Cardigan
it swings westwards obliquely across the trend of the Caledonian folds.
Farther north, Plynlimon itself lies on the crest of an analogue of the Teifi
anticline.

In regional relief much of mid-Wales gives the appearance of a dissected
plateau, whose regularity of outline, often gently rounded in 'moel' form in
its residual elements, is broken only by the deep clefts of the river valleys
or by glacial gouging. Where igneous rocks occur, however, bare crags
give rise to a more varied scenery, well seen, for example, in the neighbourhood of Builth and Llanwrtyd, and in Pembrokeshire (*see* Pls. IIIA,
IIIB). Igneous rocks interbedded with, or intruded into, softer sediments
have also a great influence on the coastal profile, the differential erosion
displayed, for example, along the indented Pembrokeshire coast from
Fishguard to St. Bride's Bay contrasting with the smooth sweep of Cardigan
Bay. It is almost literally true that every headland between Strumble and
Ramsey Island is of igneous rock and every bay and inlet etched in sedimentary rock (*see* Fig. 7; Pl. IIA); and on the south side of St. Bride's Bay,
the rocks of Skomer Island and Wooltack Point are also of igneous origin.

The wide expanse of gently dipping Old Red Sandstone to the north of
the South Wales coalfield is largely in Red Marls surmounted by the escarpment of the Senni Beds and Brownstones. Drained by the River Usk and its
tributaries, much of the ground is relatively low-lying, and, the Old Red
Sandstone producing a richer soil than the Ordovician and Silurian rocks,
is given over to agriculture and dairy farming more widely than the Cardiganshire hills. Even so, the heights of Mynydd Epynt, of the Black Mountains
east of Brecon, and of the Fans and the Brecon Beacons are almost as wild
and as barren as the Plynlimon country (*see* Pl. XIIA).

The coalfield itself is sharply delineated by the differential erosion of
hard and soft beds (*see* Fig. 4). The grits and conglomerates of the Upper
Old Red Sandstone, tough and resistant, form the summits—the highest in
South Wales, Pen-y-fan at 2906 ft—of the Brecon Beacons (*see* cover
photograph). Dipping gently southwards, they cap or contribute to a bold
and almost unbroken escarpment, with a steep northern face, that with a
score of peaks exceeding 2000 ft runs for 30 miles from Banau Sir Gaer
eastwards to the Black Mountains, and then at lesser heights continues
southwards by Abergavenny and Pontypool to the Ebbw and Taff valleys.
Within this major escarpment are parallel ridges rising to 2000 ft in several

summits, formed by the Carboniferous Limestone and (notably) the Mill-stone Grit, which dip under low ground along the outcrop of softer shales of the Lower and Middle Coal Measures, an outcrop drained by a number of subsequent strike streams. In their turn the shales dip beneath the thick Pennant Sandstone, which rises in a steep and forbidding scarp to heights of nearly 2000 ft in Craig-y-Llyn (*see* Pl. VIIB), and which forms within the coalfield an undulating plateau, unbroken except by narrow deeply incised

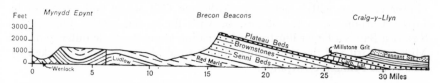

FIG. 4. *The scarped profile of the northern flank of the South Wales coalfield*

The thick sandstones of the Old Red Sandstone, the Millstone Grit, and the Pennant Measures, gently dipping southwards, form the dominant elements of the landscape.

valleys (*see* Pl. VIIA), that includes the uplands of Blaenau Morganwg— bare moorland, very thinly inhabited, given over almost entirely to the rearing of a few sheep. On the other hand, the rivers of the coalfield, cutting through the escarpment, follow courses that bear little relationship to the geological outcrops and have all the appearance of being superimposed. Only the Neath and the Tawe flow for any considerable distance along structurally determined lines.

However varied in detail, the upland regions of the Lower Palaeozoic rocks, the Old Red Sandstone, and the Coal Measures present a general similarity of aspect, not only because they consist of sandstones or toughened shales, resistant to denudation, but also because they have been affected in similar ways by the erosive events of a post-Carboniferous geological history. Although cut by such large rivers as the Towy, Wye, and Usk, they form a broad geomorphic unit that is sharply demarcated from the low country of the Vale of Glamorgan, of Gower, and of the coastal regions of Carmarthenshire and Pembrokeshire. Yet these low-lying areas also are composed mainly of folded Palaeozoic rocks, and at one time were covered by sediments several thousand feet thick that must have towered as moun-tains rivalling in height those of central Wales. The contrast in topography is thus due less to differences in rock types than to differences in erosional history: the coastal tracts have been submerged beneath the sea inter-mittently since early Mesozoic times, and are the eroded relics of low-level platforms benched by marine denudation (p. 122). In the platforms harder beds tend to stand out as residual monadnocks, like the hills of Old Red Sandstone in Gower and the Tenby district and of igneous intrusions near St. David's (*see* Pls. IIA, IIIB).

The uniform sweep of the coastline of Cardigan Bay is partly a reflection of its parallelism with the Caledonian strike of the rocks. The coastal out-line of south-west and southern Wales, where contrasted rock types and sharp folds impose differential controls, is much more varied. In south

Pembrokeshire and Gower massive Lower Carboniferous limestones tend to be the dominant elements of the headlands and the precipitous cliffs; and the lithologically similar but much younger limestones of the Lias form equally bold cliffs along much of the coast of south Glamorgan between Southerndown and Lavernock (*see* Pls. VB, VIB, XB). St. Bride's Bay reflects a syncline mainly in shales of the Coal Measures. Carmarthen Bay is eroded partly in Upper Carboniferous shales, partly in the Red Marls of the Old Red Sandstone. The alternations of bay and headland in Gower conform with synclines in Millstone Grit shales and anticlines in Lower Carboniferous limestones, or with faults and master joints; and Swansea Bay is a larger illustration of erosion along the shale outcrop of the Millstone Grit and the Lower and Middle Coal Measures, and is backed by hills of the Pennant escarpment.

History of Research

As North has shown, early geology had some of its roots in South Wales. Giraldus Cambrensis in the 12th century recognized pyritous shales near Newport; Leland in the 16th century knew the difference between the anthracites of the Gwendraeth valley and the coking coals of Llanelli; George Owen at the beginning of the 17th century was able to follow the Carboniferous Limestone at outcrop around the coalfields of Pembrokeshire and South Wales, and discerned the meaning of a geological map; Gibson before the end of the 17th century provided some of the first published illustrations of fossil plants (from the Coal Measures near Neath); and Llwyd early in the 18th century not only described fossil corals and trilobites, but was a fore-runner of William Smith in correlating the (Carboniferous) limestones of Barry with those of Caldey Island by means of crinoids that he recognized were different from the crinoids of the (Liassic) limestones of Penarth and Newport.

By the end of the 18th century William Smith was able to delineate the broad structure of the South Wales coalfield on his prototypic map by tracing the outcrop of the Red Sandstone of Brecon, the 'Derbyshire Limestone', and the Millstone Grit and Coal Measures. On the foundations he laid there was rapid advance in the early decades of the 19th century: Greenough prepared a map of Glamorgan that was astonishingly detailed and accurate for the time of its production; Conybeare described the multiple folding to be seen in the South Wales coalfield; Buckland linked prehistory with geology in his exploration of the Gower caves; and De la Beche began his systematic study of the regional geology of South Wales.

In a very few years after 1830 the major features of stratigraphy and structure in the geological frame of South Wales were established. Beyond the coalfields, which hitherto had attracted most attention, Sedgwick and Murchison, at first jointly but later in rivalry, began their studies of the Old Red Sandstone and the underlying rocks; and already by 1839 Murchison had published his *Silurian System* as the foundation for all subsequent work on Lower Palaeozoic geology. It took a further thirty years for the details of the Cambrian and Pre-Cambrian rocks of Pembrokeshire to be elucidated, mainly by Hicks (a country doctor living at St. David's), whose work allowed the rock sequence to be understood in broad outline, even if many problems of correlation and structural interpretation still remained to be solved.

The Geological Survey was founded in 1835 under De la Beche, who (greatly helped by W. E. Logan) included in its first volume of memoirs a conspectus of the geology of South Wales that set a pattern and a standard for all future work. In the mid-century years Andrew Ramsay, newly recruited to the Survey, also extended a formal geology to include the interpretation of landscape, and ascribed the formation of the Welsh plateau to uplift after marine planing. Towards the end of the century a revival of Survey interest in systematic and economic geology resulted in the sustained mapping, on the scale of six inches to one mile, of the whole of the belt of Carboniferous and associated rocks between Pembrokeshire and Monmouthshire, and in the publication of maps, sections, and a dozen memoirs that, under the general management of Aubrey Strahan, constituted for their time an impressive contribution to the understanding of a major natural region of Britain.

In the 20th century increasing attention has been given to specific problems, especially in stratigraphy. The work of O. T. Jones is of particular importance in extending the Survey discoveries northwards into the Lower Palaeozoic rocks of mid-Wales, and in providing a basis of stratigraphical interpretation both for a refined structural analysis and for a detailed palaeogeography of an evolving geosyncline. E. E. L. Dixon brought order into the stratigraphical relations of the Lower Carboniferous rocks, his pioneering work throwing a flood of light on the meaning of facies and on modes of formation of different kinds of limestone. Trueman studied the unpromising mussels of the Coal Measures, and found them to be of the greatest use in stratigraphical zoning and exemplary as indices of evolutionary mode. South Wales is now a 'type' area for a great part of the Palaeozoic rock sequence in Britain.

In recent years, a systematic resurvey by the Geological Survey of the main South Wales coalfield has resulted in the publication and prospective publication of new six inches to one mile National Grid uncoloured geological maps and one inch to one mile New Series colour-printed geological maps, and new editions of the memoirs.

References[1]

ANDERSON, J. G. C. 1960. Geology. *In* J. F. Rees (editor): *The Cardiff region*, 22–44. Cardiff.

BASSETT, D. A. 1961. *Bibliography and index of geology and allied sciences for Wales and the Welsh borders 1897–1958*. Cardiff. (Nat. Mus. Wales.)

—— 1963. *Bibliography and index of geology and allied sciences for Wales and the Welsh borders 1536–1896*. Cardiff. (Nat. Mus. Wales.)

—— 1967. *A source-book of geological, geomorphological, and soil maps for Wales and the Welsh borders*. Cardiff. (Nat. Mus. Wales.)

CHALLINOR, J. 1951. Geological research in Cardiganshire 1842–1949. *Ceredigion*, **1**, 144–76.

COX, A. H. 1920. The geology of the Cardiff district. *Proc. Geol. Assoc.*, **31**, 45–75.

---

[1]For a list of Survey memoirs, maps, and sections of South Wales, see pp. 138–9.

FEARNSIDES, W. G. 1910. North and central Wales. *Jubilee Vol. Geol. Assoc.*, 786–825.

GEORGE, T. N. 1939. *The geology, physical features, and natural resources of the Swansea district.* Cardiff.

JONES, O. T. 1930. Some episodes in the geological history of the Bristol Channel region. *Rep. Brit. Assoc.*, 57–82.

—— 1950. The structural history of England and Wales. *Rep. 18th Internat. Geol. Congr.* (Gt. Britain), **1**, 216–29.

—— 1956. The geological evolution of Wales and the adjoining regions. *Quart. J. Geol. Soc.*, **111**, 323–57.

LEACH, A. L. 1933. The geology and scenery of Tenby and the south Pembrokeshire coast. *Proc. Geol. Assoc.*, **44**, 187–216.

MITCHELL, G. F. 1962. Summer field meeting in Wales and Ireland. *Proc. Geol. Assoc.*, **73**, 197–214.

NORTH, F. J. 1928. Geological maps. Cardiff. (Nat. Mus. Wales.)

—— 1933. From Giraldus Cambrensis to the geological map. *Trans. Cardiff Nat. Soc.*, **64**, 20–97.

—— 1934. From the geological map to the Geological Survey. *Trans. Cardiff Nat. Soc.*, **65**, 42–115.

—— 1955. *The evolution of the Bristol Channel.* Cardiff. (Nat. Mus. Wales.)

—— 1955. The geological history of Brecknock. *Brycheiniog*, **1**, 9–77.

OWEN, T. R. and OTHERS. 1966. Summer (1964) field meeting in South Wales. *Proc. Geol. Assoc.*, **76**, 463–96.

SIMPSON, B. 1955. Field meeting in South Wales. *Proc. Geol. Assoc.*, **65**, 328–37.

STRAHAN, A. 1910. South Wales. *Jubilee Vol. Geol. Assoc.*, 826–58.

THOMAS, T. M. 1961. *The mineral wealth of Wales and its exploitation.* Edinburgh.

TRUEMAN, A. E. 1924. The geology of the Swansea district. *Proc. Geol. Assoc.*, **35**, 283–315.

WOOD, A. (editor). 1969. *The Pre-Cambrian and Lower Palaeozoic rocks of Wales.* Cardiff.

2. Pre-Cambrian Rocks

The oldest rocks of the region are confined to Pembrokeshire. They are the products of intense igneous activity during later Pre-Cambrian times. It is possible that they correspond in part to the Uriconian volcanic rocks of Shropshire and the Padarn and Bangor volcanic rocks of North Wales. There are few traces of an underlying platform of gneiss and schist, as in Anglesey, Lleyn, and Shropshire, but the occurrence of pebbles of quartz-schist in the Arenig conglomerates near Carmarthen suggests that such rocks may be present at depth.

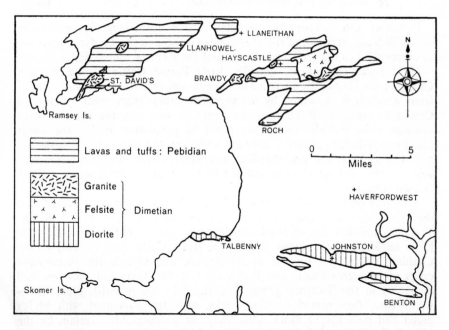

Fig. 5. *Outline map of the distribution of Pre-Cambrian outcrops in Pembrokeshire*
(In part after Thomas and Jones.)

The Pembrokeshire rocks fall into two groups, volcanic and intrusive (*see* Fig. 5). The volcanic rocks, originally described by Hicks as Pebidian, include rhyolitic and trachytic tuffs with interbedded flows of lava. The tuffs were invaded by acid plutonic intrusions, including granite, granophyre, quartz-porphyry, and quartz-diorite, that form the Dimetian suite of Hicks's classification. Both groups were folded (but not much metamorphosed), and extensively peneplaned, before deposition of the Cambrian rocks.

Pebidian

In the neighbourhood of St. David's, the Pebidian tuffs occupy a broad anticlinal inlier extending east-north-east for several miles inland. They are well displayed in the cliffs south and west of the city, where Green established the following divisions (in descending order):

		Thickness in feet
4.	Ramsey Sound Series: fine-grained sericitic tuffs – –	700
3.	Caerbwdy Series: greenish acid rocks with a quartz-chlorite matrix and with bands of halleflinta and conglomerate (Clegyr Conglomerate)	2400
2.	Treginnis Series: hard gritty rocks with abundant trachytic pumice and boulders of red keratophyre	600
1.	Penrhiw Series: gritty red and green tuffs passing down into red and green halleflinta, base not seen	1000+

In the eastern outcrops the pyroclastic rocks form horst-like masses, in which Williams recognized the Treglemais Group and the Treffynnon Group to correspond to the Caerbwdy Series and the Ramsey Sound Series.

Eight miles east of St. David's, from Pointz Castle to the Western Cleddau, the Hayscastle anticline lies in echelon with the St. David's inliers, from which it is separated by the complementary Tremaenhir syncline of Cambrian rocks. In it Thomas and Jones showed that the Pre-Cambrian volcanic suite includes two main types of pyroclastic rocks: the lower (the Pont-yr-hafod) group of tuffs are andesitic; the upper (the Rhindaston and Gignog) groups are rhyolitic and keratophyric and are associated with rhyolite and quartz-keratophyre lavas, some of which show marked fluxion structure.

Although direct proof of age is lacking it is now generally accepted that the Roch rhyolites and the associated Nant-y-Coy Beds in the Roch–Trefgarn anticline are of Pre-Cambrian age. The rhyolitic lavas and tuffs form an almost continuous ridge from Roch Castle (*see* Pl. IIB) to the Cleddau Valley and beyond to Ambleston, and include the picturesque crags of Maiden's Castle and Poll Carn (*see* Pl. IIIA); and they are well displayed in the Trefgarn gorge. The rhyolites are greenish blue, white-weathering, fine-grained silicified rocks, and are associated with bluish green and pale mauve flinty tuffs. They are conformably overlain by the Nant-y-Coy Beds, probably about 600 ft thick, flaggy tuffs considered to be the equivalents of the Ramsey Sound Series.

Separated by several miles from the outcrop about St. David's, two distinct groups of igneous rocks along the southern boundary of the Pembrokeshire coalfield emerge as faulted inliers brought to outcrop by thrusting in the Benton zone. The extrusive rocks, the Benton Series, are probably of Pebidian age. They extend from Benton Castle on the River Daucleddau to Rosemarket, and include felsites and banded and spherulitic rhyolites interbedded with fine-grained tuffs and breccias. Some of the beds may be brecciated lavas. Near Roman's Castle pink and greenish rhyolites contain large conspicuous spherulites.

(*Cambridge Univ.*)

A. St. David's Peninsula and Ramsey Island

B. The Pembrokeshire coast between Solva and Newgale

(A.6057)

(A.614

A. Maiden's Castle and Trefgarn Gorge

B. Pliocene Platform, St. David's Peninsula

(A.6099)

Dimetian

The Pebidian tuffs were invaded by acidic rocks including granite, quartz-porphyry, and quartz-diorites (*see* Fig. 5). The best known of them is the granite or alaskite-granophyre at St. David's. It is a highly siliceous coarse-grained rock, traversed by crush bands, of which the principal constituents are quartz, orthoclase, oligoclase, and chlorite. The associated quartz-porphyry is related petrographically to the granite and was probably a differentiation product of the same magma. It also is a coarse-grained rock, consisting of large phenocrysts of quartz, alkali feldspar, and a little biotite. Similar rocks are found farther east in the Llanhowel and Hayscastle outcrops.

Other igneous masses are relatively isolated, and not all of them fall into a consanguineous Dimetian suite. A small intrusion of diorite at Hollybush is a medium- to coarse-grained greenish rock, composed of quartz and biotite with large hornblende crystals. In places the quartz is present locally in sufficient abundance to make the rock a quartz-diorite, a rock showing a petrographic relationship with the quartz-diorites of the Johnston outcrop. The diorite near Knaveston, however, has no parallel: it consists of hornblende, albite-oligoclase feldspar, a little augite, and much sphene, with ilmenite and apatite as accessory minerals.

The Johnston Series, faulted against Coal Measures on the southern flank of the Pembrokeshire coalfield, occupies an area about Llangwm and Johnston, a smaller outcrop forming the cliffs facing St. Bride's Bay north of Talbenny. It includes quartz-diorites, quartz-albite rocks, and quartz-dolerite, perhaps as intrusions, perhaps as the floor on which the lavas of the Benton Series rest. The quartz-diorite is an aggregate of quartz, alkali feldspar, hornblende, and biotite, grading into a rock approaching a soda-granite. The rocks are cut by veins of quartz-albite and by dykes of quartz-dolerite up to 10 ft in width usually with chilled margins. Many of the acid rocks near Talbenny have a well-marked gneissose structure, believed to be due to post-consolidation movements. Cambrian rocks being absent, direct proof of the age of the plutonic rocks is lacking, but there seems little doubt that they are Pre-Cambrian, if not certainly Dimetian.

The Pebidian rocks are also intruded by dykes of hornblende-porphyry and dolerite, whose Pre-Cambrian age, however, has yet to be proved.

References

CLAXTON, C. W. 1963. An occurrence of regionally metamorphosed schists in south-west Pembrokeshire. *Geol. Mag.*, **100**, 219–23.

COX, A. H. and OTHERS. 1930. The geology of the St. David's district, Pembrokeshire. *Proc. Geol. Assoc.*, **41**, 241–73.

—— and THOMAS, H. H. 1924. The volcanic series of Trefgarn, Roch, and Ambleston, Pembrokeshire. *Quart. J. Geol. Soc.*, **80**, 520–48.

GREEN, J. F. N. 1908. The geological structure of the St. David's area. *Quart. J. Geol. Soc.*, **64**, 363–83.

—— 1911. The geology of the district around St. David's, Pembrokeshire. *Proc. Geol. Assoc.*, **22**, 138–41.

HICKS, H. 1877. On the Precambrian (Dimetian and Pebidian) rocks of St. David's. *Quart. J. Geol. Soc.*, **33**, 229–41.

Hicks, H. 1878. Additional notes on the Dimetian and Pebidian rocks of Pembrokeshire. *Quart. J. Geol. Soc.*, **34**, 153–63.

—— 1879. On a new group of Precambrian rocks (the Arvonian) in Pembrokeshire. *Quart. J. Geol. Soc.*, **35**, 285–94.

—— 1884. On the Precambrian rocks of Pembrokeshire, with especial reference to the St. David's district. *Quart. J. Geol. Soc.*, **40**, 507–47.

Morgan, C. Ll. 1890. The Pebidian volcanic series of St. David's. *Quart. J. Geol. Soc.*, **46**, 241–69.

Thomas, H. H. and Jones, O. T. 1912. The Precambrian and Cambrian rocks of Brawdy, Hayscastle, and Brimaston. *Quart. J. Geol. Soc.*, **68**, 374–401.

Williams, T. G. 1934. The Precambrian and Lower Palaeozoic rocks of the eastern end of the St. David's Precambrian area, Pembrokeshire. *Quart. J. Geol. Soc.*, **90**, 32–75.

3. Cambrian System

Like their equivalents in other parts of Britain, the Pre-Cambrian rocks were folded and elevated and subjected to intense erosion before the beginning of Cambrian times, unroofed Dimetian granite being exposed in places. When the region subsided beneath the advancing Cambrian sea, the first sediments were thus deposited with strong unconformity on the varied foundation beneath.

The sea appears to have remained comparatively shallow throughout Cambrian times. In the lower part of the sequence the rocks are principally sandstones with thick beds of conglomerate at the base. A slight deepening is indicated by the flaggy mudstones forming the middle part; and the banded sediments of the upper part may indicate a 'foredeep' environment. No contemporary volcanic lavas are known to occur.

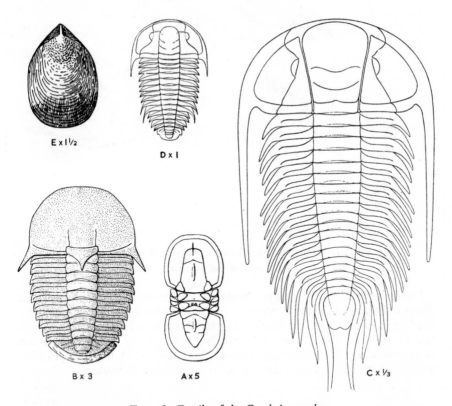

E x 1½

D x 1

B x 3

A x 5

C x ⅓

FIG. 6. *Fossils of the Cambrian rocks*

Middle Cambrian: **A.** *Tomagnostus* [*Agnostus*] *fissus* (Linnarsson); **B.** *Hartshillia inflata* (Hicks); **C.** *Paradoxides davidis* Salter. Upper Cambrian: **D.** *Olenus cataractes* Salter; **E.** *Lingulella davisii* (McCoy).

The Cambrian rocks contain the earliest Welsh records of marine life. The fossils include sponges, brachiopods, and molluscs. Trilobites are particularly important and serve as zonal indices (*see* Fig. 6). Graptolites, characteristic fossils of the succeeding Ordovician and Silurian rocks, are unknown in the region, the uppermost Cambrian series, the Tremadoc, in which they first appear elsewhere in Britain, not being represented in South Wales.

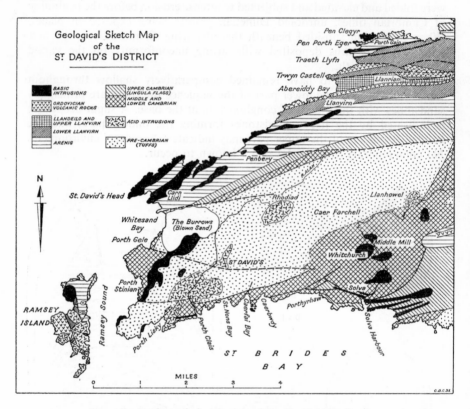

FIG. 7. *Geological sketch map of the St. David's district*
(After Cox and others.)

Cambrian outcrops, like Pre-Cambrian, are confined to Pembrokeshire. Almost the full sequence is exposed in the cliffs between Newgale and St. David's, particularly in the headland separating Caerfai and Caerbwdy bays (*see* Pl. IIB, and Fig. 7), where the beds dip steeply off the Pre-Cambrian base, and where Hicks first determined the sequence. A thickness of over 4000 ft of terrigenous sediments spans the following divisions:

*Thickness
in feet*

4.   Lower *Lingula* Flags: thin siliceous sandstones and grey shales
     with *Lingulella davisii, Olenus cataractes,* and *Homagnostus*
     [*Agnostus*] *obesus*   2000

Thickness
in feet

3. Menevian Series:
  (c) coarse grits and shales with *Billingsella* [*Orthis*] *hicksii* . 100
  (b) dark flaggy mudstones with *Paradoxides davidis* . . 350
  (a) grey flags with *Paradoxides hicksii* 300

2. Solva Series:
  (c) grey flags with *Paradoxides aurora* and *Bailiaspis dalmani*
  [*Conocoryphe bufo*] 150
  (b) green and purple mudstones and sandstones with *Ctenocephalus* [*Conocoryphe*] *solvensis* 250
  (a) green pebbly sandstones with *Paradoxides harknessi* and
  P. sedgwickii 150

1. Caerfai Series:
  (d) purple feldspathic sandstone (Caerbwdy Sandstone) . 250–500
  (c) red shales with ostracods 40
  (b) green fine-grained feldspathic sandstones, unfossiliferous 250–400
  (a) red conglomerate 60

Caerfai Series

The basal conglomerate has a striking appearance due to the colour of its pebbles, most composed of quartz, quartzite, and acid tuffs some of which are derived from the underlying Pebidian rocks, a few of quartz-porphyry of Dimetian origin. Near Hayscastle pebbles of granite and diorite also occur. The conglomerate rests unconformably on various members of the Pre-Cambrian suite: south of St. David's in the Caerbwdy valley it cuts across the Ramsey Sound Series and comes to rest on the Caerbwdy Series. Most of the junctions with Dimetian rocks are faulted but near Porth Clais the conglomerate lies on Dimetian granophyre.

The conglomerate is followed by fine-grained unfossiliferous green felds-pathic sandstones, rich in chlorite and epidote, that merge into the over-lying red shales. The conglomerate and the sandstones together increase in thickness from about 300 ft near St. David's to 450 ft in the eastern outcrops, where some of the strata are cross-bedded.

The red shales are conspicuous in the cliffs at Caerfai, at Castell on Ramsey Sound, and at Cwm-mawr near Newgale, the group forming a convenient marker recognizable over much of the region. The shales, which are cross-bedded in the lower part, are the earliest from which fossils have been obtained, although the fossils everywhere are rare. Hicks recorded '*Lingulella ferruginea*, *L. primaeva*, *Discina*, and *Leperditia*', and fragments of trilobites. The ostracod *Indiana*? [*Leperditia*] *cambriensis* has also been found: it is a form generally assumed to indicate a Lower Cambrian age. Some of the shales carry 'worm' burrows.

Purple and grey sandy layers near the top of the red shales display a lithological transition upwards to the highest member of the Caerfai Series, the purple Caerbwdy Sandstone, a fine-grained unfossiliferous feldspathic rock. At a high level in Caerbwdy Bay the Sandstone becomes coarser and carries small granite pebbles perhaps derived from Dimetian sources. Unlike the green sandstones beneath, the formation diminishes in thickness eastwards from about 500 ft near St. David's to about 250 ft.

Solva Series

The divisions of the Solva Series are well displayed in Solva harbour, on the west side of Caerbwdy Bay as far as Pen Pleidiau (*see* Fig. 7), and in the cliffs to the west of Ogof Llesugn. East of Caerbwdy they are introduced by a spectacular overfold. On the northern coast, north-west of Granston, rocks referred to the series by Cox include basal beds of coarse pebbly sandstones resting abruptly (perhaps with non-sequence) on the Caerbwdy Sandstone, and passing up into micaceous sandstones: the whole forms the Lower Solva division of Hicks. The rocks yield the trilobites *Condylopyge* [*Agnostus*] *cambrensis, Bailiella* [*Conocoryphe*] *lyelli, Paradoxides harknessi, P. sedgwickii* and *P. sjoegreni,* and the remains of one of the earliest sponges, *Protospongia major.* The sudden appearance of the trilobite fauna suggests that the lithological discordance at the base of the formation is a sign of major palaeogeographical change.

Evenly bedded green sandstones, purple and green mudstones, and grey flags that follow the Lower Solva Beds on the west side of Caerbwdy Bay constitute the Middle Solva division of Hicks: they are more than 250 ft thick, perhaps as much as 1500 ft thick (the sequence being much broken by strike faults). They have yielded the trilobite *Ctenocephalus solvensis.*

The Upper Solva Beds consist of about 150 ft of grey flags and grits. Fossils are rare, but *Paradoxides aurora* and *Bailiaspis dalmani* are recorded.

Menevian Series

The Upper Solva Beds are followed by over 600 ft of fine-grained grey flags and striped shales and mudstones with relatively abundant fossils—rocks well displayed at Porth-y-rhaw, where they are overlain by coarser siltstones, sandstones, and grits, about 100 ft thick, containing the brachiopod *Billingsella* [*Orthis*] *hicksii.* The *Orthis hicksii* Beds mark the commencement of conditions that prevailed during the deposition of the overlying *Lingula* Flags; and but for a record of *Paradoxides* and '*Conocoryphe*' by Hicks, proving a Middle Cambrian age, they would, on lithological grounds, be more appropriately regarded as forming the first members of the Upper Cambrian sequence.

The series is divided into two trilobite zones, a lower zone of *Paradoxides hicksii* and an upper zone of *P. davidis.* The shales of the lower zone contain, in addition to the zonal fossil, *Tomagnostus* [*Agnostus*] *fissus, T. sulcatus, Peronopsis* [*A.*] *exarata, Eodiscus scanicus,* and *Solenopleura applanata.* In the upper zone the giant trilobite *Paradoxides davidis* occurs in dark flaggy mudstones in association with *Holocephalina primordialis, Ptychagnostus* [*Agnostus*] *punctuosus, Peronopsis exarata, Eodiscus punctatus, Meneviella* [*Erinnys*] *venulosa, Hartshillia inflata, Solenopleura variolaris, Centropleura henrici,* and the shells *Acrotreta sagittalis* and *Hyolithes penultimus.* In some layers *Protospongia fenestrata* is abundant.

About Brawdy and Hayscastle the Menevian Series is represented by the coarse sandstones of the Musland Grit, about 150 ft thick, and the overlying shales of the Ford Beds (with fossils characteristic of the zone of *Paradoxides hicksii*). The Grit is sharply distinct in lithology from the Welsh Hook Beds immediately beneath, the youngest of which appears to be no younger than Lower Solva in age. Middle and Upper Solva beds are absent, and the lithological contrast is thus a sign of a notable unconformity or non-sequence and of contemporaneous uplift in Middle Cambrian times with overlap and overstep against a shore rising to the east and south (*see* Fig. 8).

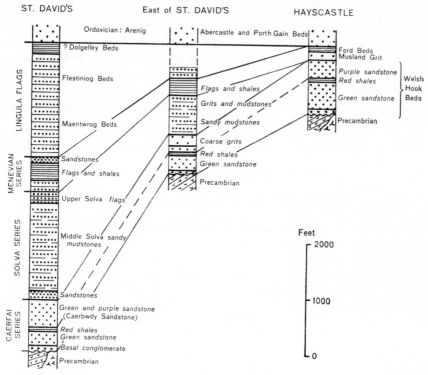

FIG. 8. *Comparative columns showing the development of Cambrian rocks in Pembrokeshire*

The signs of contemporary shores are the occurrence of thinning and overstep eastwards and south-eastwards.

(In part after Stubblefield.)

Lingula Flags

The gritty *Orthis hicksii* Beds suggest shallow-water deposition towards the close of Menevian times. They are followed by about 2000 ft of alternating micaceous flaggy shales, sandy mudstones, and siliceous sandstones, whose rhythmic banding and graded and convolute bedding are characteristic of turbidite sediments deeper-water in origin and more uniformly persistent than the grits below. In the abundance of the brachiopod *Lingulella davisii* on some of the bedding planes they are appropriately called the *Lingula* Flags and compare in lithological facies with their equivalents in North Wales.

The Flags show no marked variations in upward sequence, nor are they highly fossiliferous. Trilobites found in them at Trefgarn Bridge in the Cleddau Valley include *Homagnostus obesus, Olenus cataractes,* and *O. mundus,* which point to an Upper Cambrian age and to horizons in the lowest (Maentwrog) stage of the series. A relative crowding of lingulellids in the shales of Whitesand Bay and Ramsey Island may indicate the overlying Ffestiniog stage. The uppermost 600 ft preserved are mainly dark shales, and although they have not yet yielded diagnostic fossils they look like the Dolgelly[1] Beds of North Wales and may be of the same age.

Upper Cambrian beds of Tremadoc age are not known in South Wales.

References

Cox, A. H. and OTHERS. 1930. The geology of the St. David's district, Pembroke-shire. *Proc. Geol. Assoc.,* **41,** 241–73.

DAVIES, H. G. and DOWNIE, C. 1964. Age of the Newgale Beds. *Nature,* **203,** 71–2.

JONES, O. T. 1940. Some Lower Palaeozoic contacts in Pembrokeshire. *Geol. Mag.,* **77,** 405–9.

NICHOLAS, T. C. 1933. The age of the Ford Beds of Pembrokeshire. *Geol.Mag.,* **70,** 383–4.

PRINGLE, J. 1911. Note on the 'Lower Tremadoc' rocks of St. David's. *Geol. Mag.,* **48,** 556–9.

STUBBLEFIELD, C. J. 1956. Cambrian palaeogeography in Britain. *Rep. 20th Int. Geol. Congr.* (Mexico), **1,** 1–48.

---

[1]This was the name originally given to the formation in 1867 and has priority over all others.—*Ed.*

4. Ordovician System

The Ordovician rocks in their general characters do not greatly differ from the Cambrian and are overlain by the similar Silurian rocks in a sequence of marine sediments that attained a cumulative thickness in mid-Wales of perhaps 38 000 ft. There was a corresponding crustal downwarp on the same scale to accommodate the prism of sediments. Its fluctuating borders can be identified in south-eastern and southern Wales on one flank, and in north-western Wales and an inferred 'Irish Sea land-mass' on the other. This major downwarp, as revealed in its sediments, is the Welsh geosyncline. The limited exposures make it poorly recognizable in the Cambrian rocks of South Wales; but it is very well expressed in the wide outcrops of Ordovician and Lower Silurian rocks, whose variable characters are indeed the main basis on which its Palaeozoic form can be reconstructed.

A temporary halt in the sagging of the geosynclinal floor, with slight folding, elevation, uplift, and erosion are indicated by the uncomformity (Tremadoc Beds absent) that lies at the Ordovician base in Pembrokeshire; but it is also more directly recognizable in the visible banking of basal Ordovician sandstones against eroded *Lingula* Flags at Whitesand Bay and on Ramsey Island. Details of the contact include undercut ledges and overhanging low cliffs in the Cambrian rocks, and coarse boulder beds of the contemporary Ordovician beach. The stratigraphical break, now to be seen in Pembrokeshire in only a few small outcrops, is of widespread development in North Wales.

The Ordovician system is divided, mainly by means of the abundant graptolites (*see* Fig. 9), into four major series—the Arenig at the base, followed by the Llanvirn, the Llandeilo, and the Bala (which is commonly split into Caradoc and Ashgill). The Arenig rocks contain the four-branched *Tetragraptus*, together with early didymograptids of extensiform shape (*Didymograptus extensus, D. hirundo*); the Llanvirn, the tuning-fork graptolites (*Didymograptus bifidus, D. murchisoni*); the Llandeilo, leptograptids and early diplograptids (*Glyptograptus teretiusculus*); and the Bala, later leptograptids together with many diplograptids (*Leptograptus flaccidus, Nemagraptus gracilis, Climacograptus peltifer, Amplexograptus arctus, Diplograptus multidens, Orthograptus vulgatus, Climacograptus wilsoni, Dicranograptus clingani*, and *Pleurograptus linearis* in the Caradoc stage of the Bala; *Dicellograptus complanatus* and *D. anceps* in the Ashgill stage of the Bala).

The graptolites are generally to be found in dark grey or black pyritous and carbonaceous shales suggesting conditions of deposition under anaerobic or euxinic conditions—deposits in which shelly benthic fossils (brachiopods and trilobites) are rare or wanting. Conversely, contemporary neritic rocks laid down in refreshed well-aerated waters, with many kinds of shelly fossils, generally contain few graptolites. The Ordovician sediments thus fall into three main facies: a graptolite shale facies, usually pelagic and off-

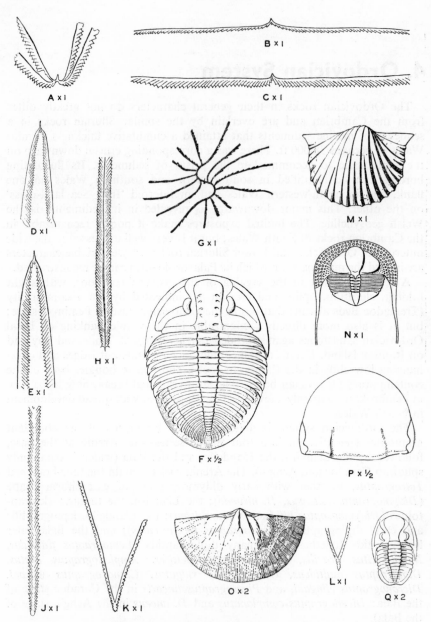

FIG. 9. *Fossils of the Ordovician rocks*
(Graptolite drawings after Elles and Wood.)

Arenig Series: A. *Tetragraptus serra* (Brongniart); **B.** *Didymograptus extensus* (Hall); **C.** *Didymograptus hirundo* Salter. **Llanvirn Series: D.** *Didymograptus bifidus* (Hall); **E.** *Didymograptus murchisoni* (Beck). **Llandeilo Series: F.** *Ogygiocaris [Ogygia] debuchii* (Brongniart). **Caradoc Series: G.** *Nemagraptus gracilis* (Hall); **H.** *Diplograptus [Mesograptus] multidens* Elles and Wood; **J.** *Orthograptus truncatus* (Lapworth); **K.** *Dicranograptus brevicaulis* Elles and Wood; **L.** *Dicranograptus clingani* Carruthers; **M.** *Nicolella actoniae* (J. de C. Sowerby); **N.** '*Cryptolithus [Trinucleus] concentricus*' auctt. **Ashgill Series: O.** *Sowerbyella sladensis* O. T. Jones; **P.** cephalon of *Illaenus bowmanni* Salter; **Q.** *Phillipsinella parabola* (Barrande).

shore; a mixed shelly, sandy, muddy, and calcareous facies, relatively near-shore; and a facies of 'poured-in' detritus, often carried by turbidity currents in accumulations of great thickness, in which ill-sorted greywackes are common, and fossils (mainly graptolites) are concentrated in intercalated shale bands. (*See* Fig. 10.)

The Ordovician period, unlike the Cambrian or the Silurian, was a time of violent igneous activity, particularly in Arenig and Llanvirn times. Lavas to great thicknesses were poured out of highly localized volcanoes, whose influence on sedimentary facies and sequence was strong in their immediate neighbourhood but diminished rapidly beyond a few miles. At origin most of the volcanoes were submarine, but their cones and vents reached heights to form islands above sea-level, and in the shallow waters around their flanks there accumulated sublittoral and beach deposits whose detritus was derived in great part from erosion of the exposed volcanic peaks or from falls of ash and tuff, and whose neritic environment encouraged a local abundance of shelly organisms. Igneous activity included intrusive phases when thick sills of dolerite were emplaced, especially in the Builth neighbourhood and in northern Pembrokeshire.

The Ordovician rocks form wide outcrops on the flanks of the St. David's and Hayscastle anticlines, and continue eastwards across Pembrokeshire into Cardiganshire as the broad Teifi anticline, and more spectacularly into Carmarthenshire and Radnorshire in the core of the elongate Towy anticline. They are hidden by Silurian rocks over much of mid-Wales, but they reappear in northern Cardiganshire in complex anticlinal inliers about Plynlimon. Narrow periclinal outcrops are also to be seen amongst Silurian and Upper Palaeozoic rocks in the cores of the Freshwater East and Castlemartin–Corse anticlines of south Pembrokeshire.

Arenig Series

The Arenig sediments comprise grits, sandstones, sandy mudstones, and grey shales. In general, the higher beds are finer-grained than the lower, and suggest a progressive deepening of the basin of sedimentation. On Ramsey Island the basal conglomerate (which contains *Bolopora undosa*, a characteristic fossil also of the equivalent horizon in North Wales) is followed by grey sandy mudstones, calcareous and highly fossiliferous, yielding a typically neritic fauna of brachiopods (*Lenorthis* [*Orthis*] *proava* and *Orthambonites* [*Orthis*] *menapiae*), trilobites (*Ogygiocaris selwyni, Synhomalonotus Calymene tristani*), and molluscs (*Palaearca*, orthocone nautiloids), with which are also found early crinoids (*Ramseyocrinus* [*Dendrocrinus*] *cambrensis*) and starfishes (*Uranaster ramseyensis*). These strata, called the Abercastle and Porth Gain Beds north of St. David's and the Brunel Beds near Trefgarn, may be followed towards Carmarthen, where as the *Peltura punctata* Beds and the Allt Cystanog Grits they become increasingly coarse-grained and conglomeratic and suggest the marginal deposits of the basin laid down at no great distance from the source area of the pebbles (of rhyolite, jasper, schist, and quartzite of Pre-Cambrian aspect) found in the conglomerates. The interbedded mudstones contain many trilobites, including *Peltura punctata, Parabolinella rugosa, Ogygiocaris selwyni*, and *Cyclopyge*.

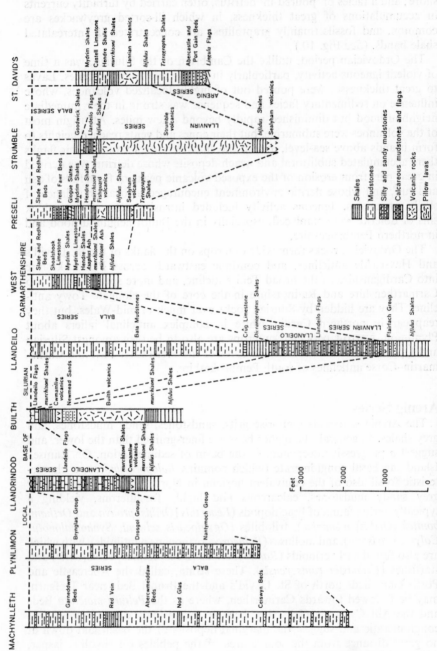

FIG. 10. *Comparative columns showing variation in thickness and lithology of the Ordovician rocks*

The upper part of the Arenig sequence is a uniform group of rusty-weathering blue and grey graptolite shales, the *Tetragraptus* Shales. They extend eastwards from the St. David's district to and beyond Carmarthen, in a thickness of about 1000 ft, without much change in character; but towards their base in the ground north-east of Fishguard the incursion of a grey-wacke suite, presumably derived from a western source, is a feature unknown elsewhere in South Wales. The beds of shale contain abundant graptolites— *Tetragraptus bigsbyi, Didymograptus extensus, D. hirundo, D. nitidus,* the early diplograptid *Glyptograptus dentatus,* the precocious single-branched *Azygograptus,* and the many-branched dendroid forms *Callograptus salteri, Dendrograptus flexuosus,* and *Dictyonema irregulare.* Occasional trilobites also occur, including cyclopygids and trinucleids.

Volcanoes were active both early and late in Arenig times. Lavas and tuffs are interbedded with sediments in the lower part of the series, some of the submarine flows, as in the Abercastle district, containing incorporated masses of mud picked up as the lavas moved over the sea-floor. The principal centres of eruption appear to have lain towards the west, beyond the present coast; but andesitic and rhyolitic lavas and tuffs are also known at low horizons in the ground between St. Clears and Carmarthen. The highly localized Trefgarn volcanic rocks, reaching at maximum thickness about 500 ft, are quarried extensively in the valley of the Western Cleddau. They are mainly composed of keratophyric flows and tuffs, with interbedded coarse conglomerates (as signs of contemporary wave erosion) composed almost exclusively of pebbles and boulders of lava. At the top of the *Tetragraptus* Shales (or perhaps at the base of the overlying Llanvirn shales) the equally localized Sealyham volcanic rocks, about 200 ft thick, in north Pembrokeshire between Abercastle and Wolf's Castle are mainly vesicular soda-rhyolites.

The most impressive development of volcanic rocks, however, is the Skomer suite of H. H. Thomas, named from its outcrops on Skomer Island. The suite can be traced for 26 miles from St. Ishmael's near Milford Haven westwards through Skomer to Grassholm Island and The Smalls. On Skomer the rocks consist of a succession of flows interbedded with bands of conglomerate, green clays, marls, and quartzites, but tuffs and intrusive rocks are rare. Neither top nor base is seen; nevertheless the rocks exposed reach a thickness of nearly 3000 ft, and the contemporary volcanic cones must have reached heights far above sea-level. On the mainland, where the suite is overlain unconformably by Upper Llandovery strata, it is much thinner and less varied, and the lavas perhaps correspond only to the two lowest members on Skomer. The flows have slaggy vesicular surfaces and show marked fluxion structures. Rock-types range from basic to acid and include soda-rhyolite, felsite, albite-trachyte, keratophyre, skomerite, marloesite, mugearite, olivine-basalt, and olivine-dolerite.

Llanvirn Series

Dark grey and blue rusty-weathering shales follow without marked difference the *Tetragraptus* Shales, and Hicks in creating the Llanvirn Series distinguished it from the Arenig on the basis not of its lithology but of its graptolites and trilobites. Its index members, tuning-fork graptolites, fall into the *Didymograptus bifidus* Zone below and the *D. murchisoni* Zone above. In the lower zone there also occur such smaller tuning-fork forms as *D. artus*, *D. nanus*, and *D. stabilis*. The upper zone yields *Didymograptus acutidens*, *D. geminus*, *Climacograptus scharenbergi*, and *Phyllograptus anna*. Trilobites, of common occurrence, include *Dalmanitina* [*Phacops*] *llanvirnensis*, *Illaenus hughesi*, *Placoparia cambrensis*, and cryptolithids. In the upper part of the zone *Cyclopyge binodosa*, *Ampyx nasutus*, *Barrandia homfrayi*, and *Illaenus perovalis* make their entry, and the shales are also marked by the incoming of *Ogygiocaris* [*Ogygia*] *debuchii*, a typical species of the overlying Llandeilo Flags.

The Llanvirn shales are widely developed in northern Pembrokeshire, where they are notably fossiliferous in Abereiddy Bay. They extend eastwards and north-eastwards into the heart of mid-Wales, appearing in periclinal cores along the Towy folds as far as the Builth inlier. In this wide development there is comparatively little lateral change in the regional lithology, except in the neighbourhood of contemporary volcanoes. Thicknesses, however, vary markedly: the *bifidus* Beds are about 1000 ft in north-west Pembrokeshire, but increase to 2000 ft near St. Clears and Carmarthen, and are of the order of 3000 ft at Builth; the *murchisoni* Beds may locally reach 500 ft, but they are usually much less and may be absent, in part because of great topographical changes in the neighbourhood of the major volcanic centres, in part because of minor warping and erosion in pre-Llandeilo times and overstep by the Llandeilo Flags. There is also some facies variation in the trilobite faunas, *Placoparia*, *Illaenus*, and *Cyclopyge*, common in Pembrokeshire, not being found at Builth, where on the other hand *Ogyginus corndensis* and *Cryptolithus gibbosus* are abundant.

Volcanic Rocks of Llanvirn Age

Extrusive igneous rocks, flows and ashes, reach a maximum development in South Wales in the Llanvirn Series. Some of them—the *murchisoni* Ash in the middle of the series, the *Asaphus* Ash immediately above it—are widespread and form useful marker bands in the stratigraphical sequence. The majority, however, tend to be thick, sometimes very thick, in the immediate neighbourhood of the extrusive centres but rapidly to be reduced laterally to insignificance. Moreover, there was a phasing in the times of maximum activity, so that volcanic outbursts and flows are not readily correlated from one locality to another, and may be misleading guides to a unified stratigraphy.

In westernmost Pembrokeshire the Llanrian volcanic suite expands from about 500 ft thick on the mainland to 1400 ft on Ramsey Island, where it is a sequence of rhyolitic ashes, tuffs, and lavas lying towards or at the top of the *bifidus* Zone. Coarse breccias and agglomerates may indicate the sites of some of the volcanic vents; and conglomerates of rounded boulders of lava are sublittoral deposits of island coasts. The Llanrian rocks are now isolated in outcrop, but originally they may have merged with the rocks (the Fishguard volcanic suite) running eastwards from Strumble Head into Mynydd Preseli, which reach a thickness of 3600 ft in the Pen Caer peninsula. In this great lenticle of lavas, three major groups have been described by Cox and by Thomas and Thomas: a lower group of rhyolites, about 800 ft at maximum near Goodwick but diminishing to 300 ft at Fishguard; a middle group of spilites, 3000 ft at Strumble but dying away east of Fishguard; and an upper group of rhyolites, 1600 ft at Goodwick, 400 ft at Fishguard, and directly succeeding the lower rhyolites in a combined thickness of as little as 300 ft locally in Mynydd Preseli. The rocks at such thicknesses emerged as islands in the Ordovician sea, and were eroded during intervals of volcanic quiescence to form pebble-beds, feldspar sands, silty mudstones, and shales, many of them ashy, which display cross-bedding and repeated non-sequence as signs of their shallow-water origin. Perhaps in consequence of the igneous restlessness, *murchisoni* Shales are absent, the Llandeilo Series resting with unconformity either on *bifidus* Beds or on the igneous rocks. (*See* Fig. 11.)

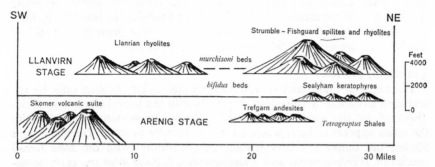

FIG. 11. *Diagrammatic representation of the main episodes of Ordovician vulcanicity in Pembrokeshire, localized in space and time*

Along the Towy fold belt, minor vulcanicity of later Llanvirn age is represented near Llandeilo by several hundred feet of rhyolitic ashes and lavas interbedded with grits and sands (called by Williams the Ffairfach Group). The sediments, some of them quartzose and arkosic, some conglomeratic with pebbles of rhyolite and keratophyre, pass laterally southwards into normal Llanvirn shales, while northwards they become thicker (like the lavas and ashes) and coarser and probably continued into the Builth district. Williams has therefore suggested that the source of both the igneous and much of the sedimentary rocks lay in a small volcanic centre between Llandeilo and Builth, the peaks of which were built up to heights above contemporary sea-level. Abundant fossils in the sands are of neritic shelly facies, like the fossils of the later Llandeilo Series (to which the Ffairfach Group was formerly referred): they include the brachiopods *Sowerbyella antiqua, Dalmanella parva, Horderleyella convexa, Hesperorthis dynevorensis, Rafinesquina llandeiloensis,* and the trilobites *Flexicalymene cambrensis, Basilicus* [*Asaphus*] *tyrannus,* and trinucleids.

The Builth volcanic rocks and the associated sediments, as interpreted by Jones and Pugh, consist of a mixed series of lavas, ashes, and agglomerates, ranging from keratophyric to spilitic types, overlain unconformably by detritus eroded from them. The volcanic rocks reach a full thickness of nearly 2000 ft, but laterally they are diminished, partly by internal reduction, partly by intermittent contemporary erosion, to only a few hundred feet, and locally they may be insignificantly developed. The peaks and islands formed by the volcanic piles were subaerially degraded, with the development of trap features and a stepped profile, into residual hills against which, on renewed subsidence, the littoral and sublittoral sediments of the Newmead Beds were deposited in progressive overlap. The sediments, many of them richly fossiliferous, range from beach gravels to feldspar sands and silty mudstones, whose contacts with the eroded volcanic rocks display in well-preserved detail such coastal features as wave-planed platforms, cliffs, notched and undercut ledges, islets, stacks, and meres and small bays. A final minor volcanic impulse is recorded in the Cwm-amliw rhyolitic ashes, which rest upon and overlap the Newmead Beds. The rocks of the Builth suite overlie lower *murchisoni* Shales, and at Llandrindod the Cwm-amliw ashes are overlain by higher *murchisoni* Shales: the whole composite episode of volcanic extrusion, degradation, and subsidence beneath the late-Llanvirn sea thus took place within the confines of a single zone, and is a relative measure of the rapid rate of geological change.

Llandeilo Series

The regional uniformity of Llanvirn sedimentation, broken only by the scattered irruption of volcanoes, contrasts with the lateral changes in facies displayed by the succeeding Llandeilo Series. In western Pembrokeshire the series appears to be represented by the Hendre or lower *Dicranograptus* Shales, not unlike the Llanvirn shales, containing (with the zone fossil) species of *Dicranograptus* and *Dicellograptus*, *Leptograptus sp.*, early nemagraptids and *Climacograptus scharenbergi*, and the trilobites *Ogygiocaris debuchii* and *Marrolithus [Cryptolithus] favus*. They are about 300 ft thick, and although not truly deep-water in origin they are in pelagic graptolite-shale facies. Eastwards they become increasingly calcareous and silty by the intercalation of fine sandstones and impure limestones, and pass laterally into the facies of Murchison's Llandeilo Flags, transitional or dominantly neritic in many beds, often with an abundance of shelly fossils of which *Ogygiocaris debuchii*, *Basilicus [Asaphus] tyrannus*, *Flexicalymene [Calymene] cambrensis*, *Trinucleus fimbriatus*, *Marrolithus favus*, *Ampyx nudus*, *Dalmanella rankini*, and *Rafinesquina llandeiloensis* are common forms. The shales, however, continue to carry graptolites, including glyptograptids, dicellograptids, early nemagraptids, *Climacograptus scharenbergi*, and *Amplexograptus perexcavatus;* and the complex interrelations of the microfacies, as Elles showed at Builth, make precise correlation difficult when subtle palaeoecological controls resulted in contrasted but laterally merging fossil suites. Thicknesses increase in the outcrops along the Towy anticline to more than 2000 ft between Llandeilo and Builth.

Minor earth-movement between Llanvirn and Llandeilo times is demonstrated in non-sequence at the base of the Llandeilo Flags, notably near Fishguard, where the shelly facies is anomalous in occurring so far north, and at Marloes, where the Flags rest on *bifidus* Beds.

Bala Series (Caradoc)

There is no sharp lithological contrast between Llandeilo and Caradoc beds over much of the outcrop from west Pembrokeshire to the north-east limits of the Towy anticline. The Hendre (lower *Dicranograptus*) Shales are followed by the Mydrim (upper *Dicranograptus*) Shales in the western development, a convenient separation of the two being made at the base of the Mydrim Limestone (the Castell Limestone of the St. David's district), which contains *Nemagraptus gracilis*. The Mydrim Shales and their equivalents to the east are highly fossiliferous, and have yielded *Dicellograptus spp.*, *Glyptograptus teretiusculus*, *Orthograptus truncatus*, *O. calcaratus*, *Diplograptus multidens*, *Climacograptus bicornis*, *Amplexograptus perexcavatus*, and *A. arctus;* but although in such a richness they are clearly rocks of the graptolitic facies, it is doubtful if they were truly deep-sea in origin, for calcareous beds with shelly fossils are found not only at their base in the Mydrim Limestone (which yields plectambonitids and trinucleids), but also towards their top in the Robeston Wathen Limestone, a typical 'Bala limestone' with neritic brachiopods (*Nicolella actoniae*, *Sowerbyella sericea*, leptaenids), trilobites ('*Asaphus*' *powisi*, phacopids, illaenids, *Encrinurus spp.*, *Acidaspis sp.*, calymenids, *Ampyx spp.*, trinucleids), gastropods, nautiloids (lituitids, orthocones, phragmoceratids), and corals (*Halysites spp.*, *Favosites spp.*, *Heliolites spp.*, '*Petraia*' *spp.*, and *Omphyma sp.*), some of which indicate very shallow depths of formation. The Crûg and Birdshill limestones near Llandeilo, coarsely crinoidal with many fossils, are massive local equivalents, about 80 ft thick, of the Robeston Wathen Limestone.

There is a merging upwards of Llandeilo into Caradoc strata along the east flank of the Towy anticline; and where they are best seen, in the neighbourhood of Builth, the lower members of the beds with *Nemagraptus gracilis* (an index fossil usually taken to define the Caradoc base) are flaggy shales and fall into the lithological group of the Llandeilo Flags. In upward sequence the beds become dominated by shales, and compare with the *Dicranograptus* Shales of the south-western outcrops; but they are much thicker, and contain many calcareous beds rich in neritic fossils interbedded with the graptolitic layers.

Volcanic rocks near Llanwrtyd found in the stratigraphical position of the Mydrim Limestone and the lower part of the Mydrim Shales are the only notable signs of Bala extrusive activity known in South Wales. They form an oval outcrop in the core of the Towy anticline, and consist of coarse breccias and tuffs with interbedded flows of spilitic lavas showing well-developed pillow-structure, the pillows being of large size and highly vesicular. The lower breccias are separated from the lavas by shales yielding Caradoc graptolites; and between the spilites and the upper ashes are bands of hornstone, tuffs, and grits and an ashy limestone with brachiopods and crinoids. The upper ashes, in which are beds of shale with *Dicellograptus sextans*, are followed by a great thickness of dark slates, much cleaved and contorted. The rocks may be regarded as a southern tongue of the Bala volcanic suite of North Wales.

Bala Series (Ashgill)

Because of the emergence of Bala rocks as anticlinal inliers in the Plynlimon country, in the heart of what is mainly Silurian terrain, the Ashgill provides the most completely known of the Ordovician contrasts between geosynclinal and shelf sediments in South Wales. In the northern outcrops from Machynlleth to Llanidloes, the dominant rocks are mudstones, with which are interbedded shales, flagstones, grits, and conglomerates, many of them greywackes showing graded, slumped, and convoluted bedding typical of 'foredeep' turbidite sedimentation. They reach thicknesses of the order of 5000 ft, and represent a major sagging of the geosynclinal floor. At Machynlleth the Abercwmeiddaw Beds, 1600 ft of pale and mottled rocks with *Dicellograptus anceps* and *Orthograptus truncatus*, underlie the Garnedd-wen Beds, 3000 ft of dark blue rusty-weathering mostly unfossiliferous mudstones. The base is not seen in the Plynlimon anticline, where O. T. Jones has recognized a lower Nant-y-moch Group, over 1000 ft thick, approximately equating with the Abercwmeiddaw Beds and containing *Dicellograptus anceps*, a middle Drosgol Group with beds of grit and conglomerate, and an upper Bryn-glas Group mainly of mudstones, the combined thickness of the middle and upper groups, together equivalent to the Garnedd-wen Beds, being about 2400 ft.

Where the Ashgill rocks reappear from beneath the Central Wales syncline, south-westwards between Llangranog and Cardigan and southwards along the Towy anticline, they are similarly thick, perhaps reaching 7000 ft about Llandrindod, and contain beds of grit, greywacke, and conglomerate in geosynclinal facies. Still farther south, however, there is relatively rapid transition to the shelly nearer-shore sediments of the margins of the geosynclinal basin. Thus in the Preseli outcrops Evans has recorded above the Mydrim Shales the cleaved Glogue Slates with *Dicellograptus anceps*, followed by the sandy mudstones, sandstones, and grits of the Freni Fawr Beds, which yield a shelly fauna with orthids, *Leptaena*, corals (*Halysites*), and bryozoans. Near Llandeilo, Williams has listed, from impure nodular limestones and calcareous sandstones and grits interbedded with mudstones, *Sowerbyella sladensis*, *Glyptorthis balclatchiensis*, rafinesquinids, leptaenids, *Skenidioides lewisi*, *Diacalymene marginata*, *Tretaspis kiaeri*, *Cybeloides sp.*, *Favosites fibrosa*, crinoids, bryozoans, gastropods, cystids, and tentaculitids. At Abbey-cwm-hir the exposed Ashgill sequence begins in the Carmel Beds, a group of shales and mudstones with some grits containing a graptolitic fauna, overlain by 1600 ft of Camlo Hill Beds in which brachiopods (including *Stropheodonta hirnantensis* at high horizons) and trilobites are like those found near Llandeilo.

The Ashgill rocks in Carmarthenshire and Pembrokeshire, from Llandeilo to Haverfordwest, are mainly neritic and shelly in a thickness approaching 2000 ft. At the base, the Shoalshook Limestone, about 200 ft of calcareous mudstones with impure nodular limestones and bands of sandy rottenstone showing much lateral variation, has abundant fossils including many cystids (*Echinosphaerites*, *Hemicosmites*, *Sphaeronites*) and brachiopods, and an abundance of trilobites (calymenids, phacopids, illaenids, cheirurids, lichids, *Phillipsinella*, *Sphaerexochus*, *Remopleurides*, trinucleids). The limestone is succeeded by the Redhill Beds, 1200 ft of olive and grey mudstones with some thin sandstones, and the Slade Beds, 300 ft of fossiliferous red-stained mudstones with *Sowerbyella sladensis* and *Chasmops conophthalmus*. Locally the Slade and Redhill Beds rest with overstep on the Mydrim Shales, an interbedded conglomerate near Llawhaden containing pebbles with Caradoc graptolites: a migratory contemporary coast not far to the south is implied.

Intrusive Rocks of Ordovician Age

Numerous sill-like dolerite intrusions are emplaced in the Arenig and Llanvirn sediments of Pembrokeshire. They range in texture from fine-grained to coarsely crystalline, and show wide variation in silica content, but are nevertheless intimately related and appear to have been derived from a common magma. The type near St. David's is a quartz-enstatite-dolerite, but the closely allied rocks of Strumble Head are quartz-free. The thick sills forming Carn Llidi and St. David's Head, described by Roach, are coarsely crystalline quartz-enstatite gabbros, products of multiple injection and differentiation: they often show ophitic structure, but where quartz is a relatively abundant constituent granophyric intergrowths occur. The doleritic masses of Carn Ysgubor, Pen Beri, and Porth Gain are medium- to fine-grained, and are more acid than the typical dolerites of the district (*see* Pl. IIIB). In Mynydd Preseli the dolerites are characterized by the presence of pink and white spots, and form a striking rock type. The Llanvirn intrusions are mainly acid rocks like the contemporary lavas. Sills of quartz-albite porphyry on Ramsey Island, occurring at a constant horizon in the *bifidus* Shales, appear to have no counterpart on the mainland: they are fine-grained, with phenocrysts of quartz and albite set in a felsitic ground-mass. The quartz-porphyry sill east of St. Clears contains large oligoclase and orthoclase crystals with small hornblende. The megalithic 'bluestones' of Stonehenge were, as H. H. Thomas showed, transported from the Preseli outcrops of dolerite.

Many intrusions of albite-dolerite occur in the Builth inlier, preferred contexts being shale groups ranging from early Llanvirn to Caradoc in age. A number appear as sills, especially at lower horizons, but many at higher horizons are smaller, discontinuous, and plug-like, and as reconstructed by Jones and Pugh they may have the bulk form of a multiple laccolith with feeders leading to minor apophyses projecting from the major sheets. The emplacement of some of the dolerites in the Caradoc sediments proves igneous activity at a later time than any known extrusives in the area, under a cover little more than 1000 ft thick of rocks presumably not as young as Ashgill; and the possibility is not to be overlooked that late-Caradoc flows were extruded but were later erosively stripped before the onset of Ashgill sedimentation.

[For references on the Ordovician rocks see pp. 47–50.]

5. Silurian System

The conditions controlling sedimentation at the close of Ordovician times persisted during early Silurian times without marked change. The major trough of the geosyncline continued to lie in western mid-Wales, where a prism of shales, mudstones, grits, and greywackes accumulated almost uninterruptedly to thicknesses probably exceeding 17 000 ft. To the east and the south the trough was flanked by marginal shelves where calcareous neritic sediments, 5000 ft or less in thickness, disclose repeated non-sequences and unconformities, a proximity to nearby migratory shores, and the effects of pulses of earth-movement and warping. The latest Silurian rocks show signs of diminishing marine influence, and there is upward passage in a lithological transition and in a restriction of the range of fossils to the continental facies characterizing the Old Red Sandstone. The close of Silurian times saw a radical change in the geological context of South Wales—an end of the Lower Palaeozoic era of sustained geosynclinal deposition that had lasted some 200 million years, and a beginning of an Upper Palaeozoic era when the frame of a prototypic Wales was first blocked out.

Silurian rocks are widely distributed in South Wales. They form the greater part of the moorlands north-west of the Towy anticline, where only small Ordovician inliers interrupt the continuity of their outcrops, and where they are well displayed along the coast from Aberystwyth to Llangranog. Over almost the whole of this large area they are in graptolitic geosynclinal facies. They also run in a narrowing outcrop (being progressively overstepped by the Old Red Sandstone) south-westwards from the Builth Ordovician inlier along the south-east flank of the Towy anticline; and they are again to be seen in isolated outliers between Whitland and Haverfordwest and in the anticlinal cores of south-west Pembrokeshire. Along this broken belt they are mainly in shelly neritic facies. It is probable that they underlie the greater part of the coalfield, for in Glamorgan upper Silurian rocks in 'normal' shelf facies are brought to the surface in the core of the Rumney anticlinal fold near Cardiff.

Thin lavas at Marloes Bay and Wooltack in Pembrokeshire are the only contemporaneous igneous rocks known.

There are no abrupt lithological contrasts between the Ordovician and the Silurian geosynclinal rocks, but important faunal changes serve to define the junction, notably the disappearance, a few diplograptids excepted, of the many graptolite genera characterizing the Ordovician rocks, and their replacement in Silurian times by species of a single group, the monograptids, whose evolutionary stages form the basis of a zonal scheme of classification (see Fig. 12). The system is divided into a lower part, the Llandovery Series, in which monograptids are accompanied by relict diplograptids, and an upper part, the Wenlock and Ludlow series (together constituting the Salopian), in which only monograptids survive (but in Wales, as elsewhere in Britain,

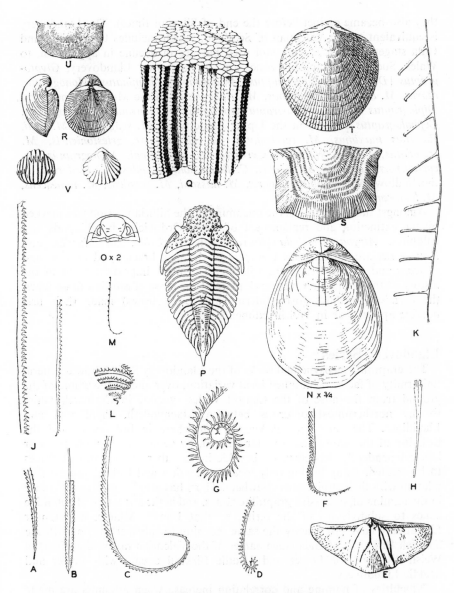

FIG. 12. *Fossils of the Silurian rocks*
(All natural size except N and O; graptolite drawings after Elles and Wood.)
Llandovery Beds: A. *Akidograptus [Cephalograptus] acuminatus* (Nicholson);
B. *Diplograptus [Mesograptus] modestus* Lapworth; C. *Monograptus cyphus*
Lapworth; D. *M. triangulatus* (Harkness); E. *Plectodonta duplicata* (J. de C.
Sowerby), internal cast; F. *Monograptus argenteus* (Nicholson); G. *M. con-
volutus* (Hisinger); H. *Cephalograptus cometa* (Geinitz); J. *Monograptus sedg-
wickii* (Portlock), distal and proximal parts; K. *Rastrites maximus* Carruthers;
L. *Monograptus turriculatus* Barrande; M. *M. crispus* Lapworth; N. *Pentamerus
oblongus* J. de C. Sowerby; O. *Phacops elegans* (Sars and Boeck); P. *Encrinurus
punctatus* (Wahlenberg). **Wenlock Beds:** Q. *Favosites gothlandicus* Lamarck;
R. *Resserella [Dalmanella] elegantula* (Dalman); S. *Leptaena rhomboidalis*
(Wilckens); T. *Atrypa reticularis* (Linné). **Ludlow Beds:** U. *Protochonetes ludloviensis* Muir-Wood [*Chonetes striatellus* Dalman sp.]; V. *Camarotoechia
nucula* (J. de C. Sowerby).

they also became extinct before the end of Silurian times). The Llandovery is equivalent to the Valentian of Scotland, which includes the Birkhill and Gala stages—terms that are still used but are in decline in application to Wales. Successive graptolite zones are: in the Lower Llandovery, *Glyptograptus* [*Diplograptus*] *persculptus, Akidograptus acuminatus, Monograptus atavus, M. acinaces, M. cyphus, M. triangulatus;* in the Middle Llandovery, *Diplograptus magnus, Monograptus leptotheca, M. convolutus, M. regularis, Cephalograptus cometa;* in the Upper Llandovery, *M. sedgwickii, M. halli, Rastrites maximus, M. turriculatus, M. crispus, M. griestoniensis, M. crenulatus;* in the Wenlock, *Cyrtograptus murchisoni, Monograptus riccartonensis, Cyrtograptus rigidus, C. linnarssoni, C. ellesi, C. lundgreni;* in the Ludlow, *Monograptus vulgaris, M. nilssoni, M. scanicus, M. tumescens, M. leintwardinensis.*

Amongst the shelly fossils the oncoming of the Silurian period was marked by the extinction and replacement of many Ordovician brachiopods and trilobites. *Atrypa, Clorinda* [*Barrandella*], *Meristina, Stricklandia,* and *Pentamerus* are conspicuous members of the new brachiopod groups, and *Phacops* and *Dalmanites* of the trilobite. The most important event in the history of the fauna, however, took place at the close of Silurian times when the earliest chordates (ostracoderms and bony fishes) made their first notable appearance in Welsh sediments.

Llandovery Series

The graptolitic geosynclinal rocks of the Llandovery Series show a general uniformity of facies, with much local variation, over almost the whole of the ground from the nose of the Central Wales syncline in Carmarthenshire to the northernmost outcrops between Machynlleth, Plynlimon, and Llanidloes. They are not yet known everywhere in full detail, and the isolation of the several areas so far studied, by geologists working more or less independently, has caused a bewildering number of formational names to be created, many having only local application and lacking precise correlation with their equivalents elsewhere. There has also been some difference in the naming of successive graptolite zones, and in the allocation of definitive zones to subdivisions of the series, so that Lower, Middle, and Upper Llandovery have not exactly the same meaning for all workers; nor do the Scottish Birkhill and Gala stage names of the Valentian match the tripartite Welsh divisions, or Lower and Middle Birkhill equate with Lower and Middle Llandovery.

Difficulties of naming and correlation increase when attempts are made to link the graptolitic facies with the shelly, especially as the shelly rocks are themselves found in fragmented outcrops. While shelly fossils—mainly brachiopods and trilobites—are almost wholly absent from the graptolitic facies, occasional marker bands with graptolites are found in the shelly facies and allow some ties to be established between the two rock suites. In general, however, the calcareous shelf sediments are divided between Lower, Middle, and Upper Llandovery by means of the brachiopods, and it is not to be assumed that correlation with the similarly classified graptolitic rocks across the Towy anticline is always precise.

Graptolitic Facies

The Llandovery rocks of mid-Wales, knowledge of which in great part is the result of work by O. T. Jones, reach their maximum development in the neighbourhood of Llanidloes and Plynlimon, where they exceed 10 000 ft in thickness (*see* Fig. 13). They are shales, mudstones, silty flags, grits, and conglomerates, with black shale bands crowded with graptolites. Many of the rocks are well bedded or laminated, but some are convoluted and slump-bedded and indicate submarine sliding. The vary greatly in colour, especially in the upper members amongst which 'green and purple' shales are conspicuous; but, as the graptolites show, the colour changes occur at different horizons in different places and are not safely to be used in correlation.

The Lower Llandovery rocks (the lower Clywedog Beds of Llanidloes, the Eisteddfa Beds and the lower Rheidol Beds of Pont Erwyd) are mud-stones followed by shales, of the order of 1800 ft thick, which have yielded the zone fossils together with *Climacograptus scalaris*, *Glyptograptus tamariscus*, *Monograptus revolutus*, and *Petalograptus palmeus*. Traced north-westwards to Machynlleth, the stage (called locally the Cwmere Group) is reduced to about 330 ft of mudstones, rusty-weathering shales, fine siliceous flags, and black graptolitic shales sometimes displaying rhythms of sedimentation. The base shows a sharp lithological contrast with the underlying Bala rocks, but there is no suggestion of unconformity. South-wards, on the other hand, the stage (as the Dyffryn Flags and overlying Ddôl Shales of the Gwastaden Group) reaches 1300 ft at Rhayader and Abergwesyn: a feature of the sequence is the development of the Cerig Gwynion Grits forming the basal beds at Rhayader, and of the Drygarn Conglomerate in the *atavus* Zone near Abergwesyn.

The Middle Llandovery rocks at Llanidloes are mainly mudstones and shales (the upper Clywedog Beds), about 500 ft thick, with many graptolite bands containing with the zone fossils climacograptids, petalograptids, *Rastrites peregrinus*, *Monograptus lobiferus*, and *M. clingani*. The stage thins to 115 ft at Machynlleth (as part of the Derwen Group) and 135 ft at Pont Erwyd (as the upper part of the Rheidol Group and the lower part of the Castell Group), but it is 400 ft thick (as the Gigrin Mudstones of the Gwastaden Group) at Rhayader, and 440 ft on the flanks of Drygarn; and it reaches 870 ft near Pumpsaint. The Ystrad Meurig Grits south of Pont Erwyd, and grits and conglomerates developed near Abergwesyn, are local intercalations interrupting the normal sequence of argillaceous and silty rocks. The *leptotheca* Band is noteworthy: it is a dark shale in which the 'green streak', an inch thick, forms a marker band of wide dis-tribution not only in Wales but in the Lake District also.

The Upper Llandovery rocks (the upper part formerly called the Tarannon Shales) reach thicknesses of about 7300 ft at maximum near Llanidloes, where an ascending sequence of Oldchapel, Caerau (Gelli), Moelfre, and Pale Shales (Dolgau) groups, described by E. M. R. Wood and W. D. V. Jones, span the *sedgwickii-turriculatus*, *crispus*, *griestoniensis*, and *crenulatus* zones. The stage is dominated by mudstones and shales, but grits occur in the *griestoniensis* Zone, and as the Talerddig Grits become increasingly important northwards. In the graptolite bands there occur with the zone

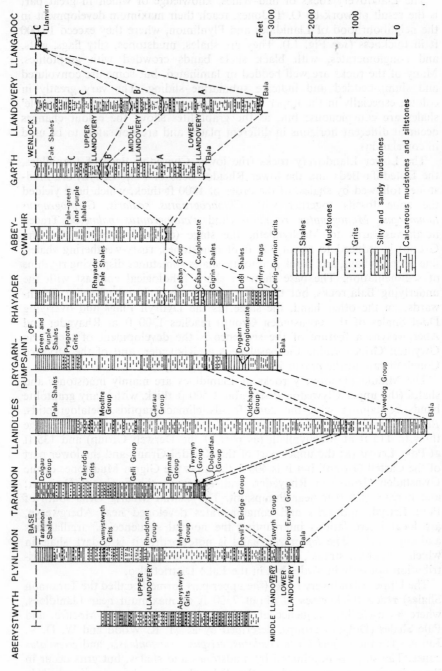

Fig. 13. *Comparative columns showing variation in thickness and lithology of the Llandovery rocks*

fossils *Monograptus marri, M. priodon*, and *Retiolites geinitzianus*. Thickness is of the same order about Pont Erwyd and perhaps at Aberystwyth, where the upper part of the Castell Group spans the *sedgwickii-maximus* zones, and the Ystwyth Group the *turriculatus-crenulatus* zones. A feature of the sequence is the development of the Aberystwyth and Cwmystwyth grits, together perhaps reaching 4000 to 5000 ft in thickness. Towards the south-east, at Rhayader where the Caban Conglomerate lies at the base and the Caban Group, 1000 ft thick, spans the *sedgwickii* and *halli* zones, the upper part of the stage, the Rhayader Pale Shales, is reduced to about 3000 ft, partly because of Wenlock overstep; and east of Abbey-cwm-hir only a thin remnant of the Pale Shales remains.

Grits in the Llandovery Series

While there is general uniformity of lithological type in the fine-grained mudstones and shales of the greater part of the Llandovery succession in mid-Wales, however much variation there may be in detail, the coarser grits and conglomerates show rapid changes in thickness and in horizon, and commonly persist for only short distances. Thus the Cerig Gwynion Grits of the *persculptus* Zone, 170 ft thick at Rhayader, are reduced to negligibly thin tongues on the flanks of Drygarn, 8 miles away: they partly overlap in range the Drygarn Conglomerate of the *atavus* Zone, a coarse pebble bed 250 ft thick that can be followed as a massive member along the strike for only some five miles before it ends abruptly in a snub-nosed outcrop. In the Middle Llandovery Stage the thick Ystrad Meurig Grits are known only around the anticlinal core about Yspytty Ystwyth. At the base of the Upper Llandovery Stage the Caban Conglomerates end bluntly south-west of Rhayader in a mile or less from outcrops where they exceed 240 ft in thickness. The thickest development of grits occurs in the Upper Llandovery rocks in which individual beds are relatively persistent but major groups are diachronous. In the wide development of the Aberystwyth Grits, the earliest members are low in the *turriculatus* Zone in southern outcrops near Newquay, but first appear perhaps as high as the *crispus* Zone near Aberystwyth. Eastwards their analogues, the Cwmystwyth and Pysgotwr grits, north-eastwards the Talerddig Grits, lie mainly in the *griestoniensis* Zone, a few thickening grit bands at Llanidloes persisting into the lowest part of the overlying *crenulatus* Zone. At higher horizons the Denbighshire Grits of the Wenlock Series are a similar formation that first appears (as the locally named Fynyddog Grits) in this ground and likewise develops northwards. The locus of deposition of thick grits thus appears to have migrated steadily across Wales as Silurian times advanced.

In lithological characters the coarser beds, some with boulders over a foot in diameter, are superficially very different from the finer quartzitic grits, flags, and sandstones of the typical 'grit' sequence; but apart from size of fragment most of the arenaceous and rudaceous rocks are united as greywackes in the poor sorting of the grains and pebbles, in the common occurrence of subangular pieces of shale amongst the pebbles of tougher rocks, and in the fine-grained argillaceous matrix. The blunt limits and the channelled form of many of the conglomeratic beds and the scattering of pebbles in a fine-grained matrix are common features of turbidite deposition. In Wood and Smith's description, the ribbon banding through many hundreds of feet of the Aberystwyth and similar grits (*see* Pl. IVA), the alternation of grits with finely laminated and gently deposited shales, the graded bedding in the grits, the occurrence of convolutions, and the disturbances on the sole of many grit bands (as load-casts and flute-casts, groove-casts and washouts) are also repeated signs of transport by turbidity currents. The regional environment was pelagic and off-shore and if not deep-water at least inimical to most benthic animals (although 'worm'-tracks and other trails are not uncommon); subsidence was relatively fast to accommodate the great thicknesses of sediment; and marginal slopes were steep to encourage the periodic triggering of turbidity slumps and flows. In a sedimentary sense, such an environment is characteristically geosynclinal, the rock-suite as a whole being 'foredeep' rather than black-shale axial in the geosynclinal frame.

Confirmation is given by the inferred sources of the detritus in the grits and conglomerates. Current-direction in the Aberystwyth Grits (as in some of the later Silurian turbidites), indicated by drag, foreset bedding, ripple-drift, diminution of grain size, and cast marks, is consistently from the south or the south-west along the trough of the geosyncline. Farther east, pebbles of igneous rocks common in the conglomerates between Rhayader and Abergwesyn suggested to Davies and Platt derivation by transverse current-transport from Ordovician and Pre-Cambrian sources to the east and south-east, where Llandovery rocks are both thin and in shelf facies, and probably form, beneath the blanket of Old Red Sandstone, a discontinuous cover on an old foundation, down to Pre-Cambrian rocks, between the Towy anticline and the Malvern ridge. (In later Silurian times current drift appears to have been mainly longitudinal, aligned with the geosynclinal trend, but north-westward sliding and slumping from the south-eastern shelf is also to be recognized, especially in Ludlow rocks.)

Transition from Geosynclinal to Shelf Facies

The outcrops of Ordovician rocks in northern Pembrokeshire and along the Towy anticline now separate the rocks of the geosynclinal and the shelf facies of the Llandovery Series. The gap is not bridged anywhere in South Wales except along a very narrow belt near Abbey-cwm-hir, where the nose of the Towy anticline plunges under the transgressive Wenlock rocks of the Clun Forest country (Fig. 14).

Signs of unrest of the Llandovery sea-floor towards the geosynclinal margin appear on the east flank of the Gwesyn syncline, where Davies found evidence of warping along north-westerly axes in an overstep of the *sedgwickii* Zone by the Rhayader Pale Shales. Farther north-east the rock relations become complex. Towards Rhayader, in Lapworth's interpretation, a thick and unbroken sequence of Lower and Middle Llandovery rocks of the Gwastaden Group was deeply eroded almost to the base, in a thickness of 1500 ft, before the transgressive deposition upon them of the Upper Llandovery Caban Group—a group displaying strong internal overlap. At

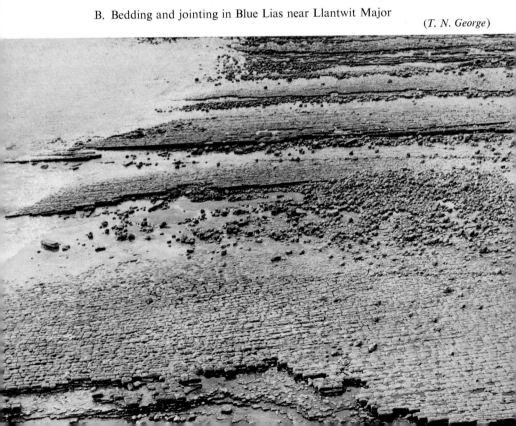

(*T. N. George*)

A. Ribbon banding and graded bedding in the Aberystwyth Grits

B. Bedding and jointing in Blue Lias near Llantwit Major

(*T. N. George*)

A. Hercynian structures in Caswell Bay, Gower   (A.108

B. 'The Green Bridge of Wales' and the South Pembrokeshire Platform   (A.10

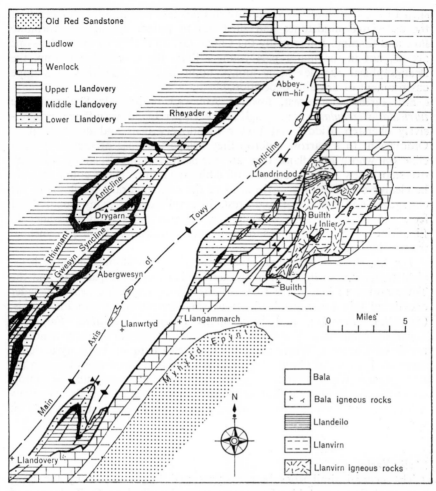

FIG. 14. *Geological map of the nose of the Towy anticline to show the development of major unconformities in the short distance across the core of the fold*

The broken outcrop of the Middle Llandovery rocks near Rhayader is a first sign of contemporary earth-movement. On the south-eastern flank there is repeated wedging-out of Lower Llandovery rocks beneath Middle, Middle Llandovery beneath Upper, and Upper Llandovery beneath Wenlock. The Builth inlier of Llanvirn rocks, affected by pre-Llandovery deformation, is surrounded by Wenlock, with only thin residual Upper Llandovery appearing patchily along its western margin.

Rhayader, in an abrupt reversal, the full Gwastaden Group returns; but it is followed directly by the Rhayader Pale Shales, the Caban Group being overstepped (*see* Fig. 13). Kelling and Woollands, however, in alternative view have ascribed the abrupt lateral changes to major slumping and turbidite channelling, not to normal unconformity. They relate the elongate multilenticular form of the Caban Conglomerates to deep gouging, in repeated phases, of a narrow submarine trench that descended from the

shelf to the east, and the variability in thickness and lithology of many of
the associated mudstones and siltstones to milder episodes of turbidite
flow and fanning. Inferentially the Caban Group is then of much the same
age as, not in the major part younger than, the Gwastaden Group, the
stratigraphical complex being looked upon as clear indication of a slope
environment of unstable deposition.

Still farther north-east, towards Abbey-cwm-hir, Pale Shales of the
turriculatus-crenulatus zones, high in the Upper Llandovery Stage, progressively overstep Middle and Lower Llandovery beds and descend on to
Ordovician. In a narrow outcrop they bridge the nose of the Towy anticline on whose east flank they are themselves extinguished by Wenlock
overstep a few miles south-east of Abbey-cwm-hir. There could scarcely
be clearer evidence of a restlessness of the contemporary sea-floor, in the
emergence and erosion of fold crests in amplitudes of many hundreds of
feet, and, across a span of only four or five miles from Rhayader to
Llandrindod, in a tectonic and sedimentary transition from geosynclinal
trough to marginal shelf.

The Llandovery shales and mudstones, richly graptolitic in many beds,
yield scarcely a shelly fossil in the terrain between Cardigan Bay and the
Towy anticline; but in the rocks of the Gwesyn syncline near Rhayader
occasional thin tongues of calcareous sandstone contain crinoids, corals,
(*Favosites*, '*Petraia*'), and brachiopods (*Atrypa*, *Meristina*, *Eospirifer*,
Leptaena, pentamerids), and are the first sign of the influence of a nearby
neritic environment. Where along the south-east flank of the Towy anticline
there are lenticles of Llandovery rocks preserved in downwarps beneath the
transgressive Wenlock rocks, the transition to the shelly facies is almost
complete, the sediments yielding an abundance of brachiopods and many
trilobites, and including only an occasional graptolite bed as a sign of
western influence.

Shelly Facies

The general similarities in rock-types and fossils shown by the now-
separated Garth and Llandovery outcrops, described by O. T. Jones and
Andrew, suggest a former continuity along a common depositional strike.
The Llandovery rocks are preserved in shallow synclines beneath the regional
Wenlock blanket, the folds running north-north-westwards across the
present-day trend of the Towy anticlinal axis and providing repeated evidence
of intra-Llandovery warping and non-sequence. Thirty miles farther west,
between Whitland and Haverfordwest, Llandovery rocks again emerge
from beneath transgressive Old Red Sandstone, and although the details
of succession are slightly changed, the facies is much like that of Llandovery
both in rock-types and in shelly fossil-suites. The main kinds of sediments
are calcareous mudstones, siltstones, and sandstones (the Llandovery
Sandstone of Murchison), with occasional beds of shale.

At Garth and Llandovery the Lower Llandovery Stage rests unconformably on Ordovician rocks. Thin grits and conglomerates locally form the
basal members, but the beds above, to a thickness reaching 2300 ft, are mainly
mudstones and fine sandstones. Amongst the fossils they yield are '*Atrypa*

marginalis', '*Meristina crassa*', *Clorinda* [*Barrandella*] *undata, Stricklandia lens, Leptaena spp., Dicoelosia* [*Bilobites*], *Platystrophia biforata,* and *Plectodonta spp.*; *Monograptus atavus* has been found in the Garth outcrop, and *M. incommodus* at Llandovery. There are signs in minor overstep of warping of the Lower Llandovery rocks before deposition of the Middle; but generally the two stages run together, the Middle being a group of mudstones, about 800 ft thick, containing (with other fossils) *Plectodonta millinensis* and *Triplesia insularis,* fairly common trilobites (phacopids, lichids, calymenids, encrinurids), some corals (*Favosites*), and a band of graptolites with *Monograptus regularis, M. lobiferus,* and *Climacograptus scalaris.*

The rocks of the Upper Llandovery Stage, mudstones and sandstones with impure nodular limestones, over 2000 ft thick in the Llandovery outcrop, carry an abundantly rich and distinctive suite of fossils including *Atrypa reticularis, Eocoelia hemispherica, E. curtisi, Eospirifer radiatus, Cyrta, exporrecta, Pentamerus oblongus, Clorinda globosa, Stricklandia lirata,* and *Strophonella euglypha,* trilobites including illaenids and cheirurids, and corals (*Favosites,* '*Petraia*'). The highest beds at Garth are pale mudstones, a last surviving outlier of the Rhayader Pale Shales, and have yielded *Monograptus crenulatus.* The stage is strongly transgressive on the beds beneath, downwarps of Lower and Middle Llandovery rocks, in amplitude exceeding 2000 ft, being recognizable both at Garth and at Llandovery, where overstep reaches an extreme in the extinction of the underlying rocks locally down to Llanvirn. It is probable that only Upper Llandovery rocks in thin and discontinuous veneer remain beneath younger formations in the shelf region to the east and south-east of the present outcrops. In a distance of little more than 6 miles across the Towy fold from the thick greywacke suite of the Mallaen–Gwesyn syncline the Llandovery sequence is reduced to insignificance, the persistently positive shelf offering major contrast in its tectonic restlessness to the constantly subsiding geosynclinal trough to the west (*see* Fig. 15).

In the Narberth and Haverfordwest outcrops, the Haverford Stage of Lower Llandovery age comprises some 2000 ft of mudstones and sandstones, with bands of conglomerate towards the unconformable base, containing shelly fossils together with a few graptolites. In ascending order, the fine mudstones of the Cartlett Beds and the sandier Gasworks Mudstones are like their equivalents at Llandovery; but the Gasworks Sandstone, with *Plectodonta superstes,* appears to lie at a higher horizon than any beds preserved at either Llandovery or Garth.

Middle Llandovery rocks are absent, because of Upper Llandovery overstep. At Haverfordwest the Millin Stage of Upper Llandovery age is dominated by mudstones (the Uzmaston and Canaston beds), about 1600 ft thick, containing *Eospirifer radiatus, Clorinda globosa,* and *Plectodonta millinensis,* and is like the rocks at Llandovery; but in inliers to the south of the coalfield the sediments become coarser and sublittoral along the margins of an inferred 'St. David's Land'. At Rosemarket the greater part of the Millin Stage is represented in the arenaceous Rosemarket Beds, which, with conglomeratic base, rest with great unconformity on Pre-Cambrian rocks, the Lower Llandovery Stage being overstepped. At Marloes and

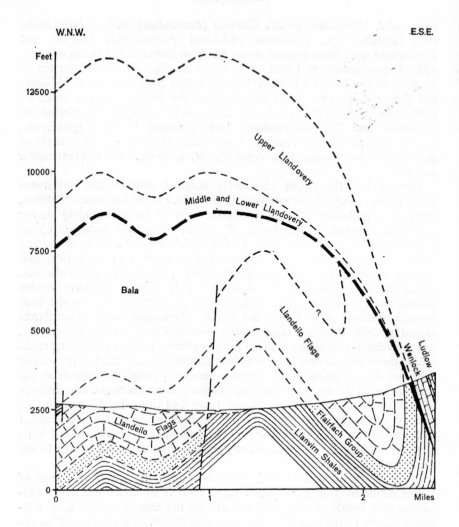

FIG. 15. *Diagrammatic reconstructed section, approximately true to scale, across the Towy anticline in the neighbourhood of Llandeilo*

Complex pre-Llandovery folding is shown by the deformation of the Llandeilo and Llanvirn rocks. The Llandovery rocks in their turn are arched into an anticline whose eastern flank shows stages of overstep beneath the Wenlock as the geosynclinal facies of the west passes laterally into the shelf facies of the east. On the western flank there is a virtually continuous sequence, including thick Bala, from Llanvirn to Wenlock. What pre-Llandovery deformation the Bala rocks suffered is conjectural. Farther north towards Llandovery pre-Wenlock folding of the Llandovery rocks was on a scale of thousands of feet in amplitude, and the diagram is probably over-simple in representing the Llandovery rocks merely as wedging out eastwards.

(After the work of Davies and Williams.)

Wooltack, conglomerates, sandstones and quartzites, and overlying mud-stones forming the lower part of the Coralliferous Series of De la Beche, yield many fossils including *Palaeocyclus porpita, Favosites spp.*, '*Petraia*' *spp.*, *Clorinda globosa, Stricklandia lirata*, and *Leptostrophia compressa*, and can be seen to be banked against Arenig igneous rocks along what is recognizably a contemporary coast. The lateral changes in sequence and sediments, south-wards in Pembrokeshire, eastwards and south-eastwards from Rhayader, Abergwesyn, and Caio to Builth, Garth, and Llandovery, thus provide a close palaeogeographical definition of the structural frame in which the Llandovery deposits accumulated.

Wenlock Series

Silurian rocks younger than Llandovery are known in the Central Wales syncline only in the small Tarannon outlier, the general absence of the Wenlock and Ludlow series being a striking testimony to the denudation the region has undergone. To the south-east, Wenlock rocks lie in a belt ranging from Newtown to Builth: they rest on a thin residue of Upper Llandovery rocks, which in places, notably around the Builth anticlinal inlier, they overstep to rest with great discordance on Ordovician rocks down to Llanvirn (*see* Fig. 14). From Builth they continue south-westwards in a narrow outcrop along the south-east flank of the Towy anticline until they are finally overstepped by the Old Red Sandstone between Llandeilo and Carmarthen. Farther west they reappear in small inliers in south-western Pembrokeshire on both flanks of Milford Haven; and far to the east they emerge at Cardiff in the Rumney inlier forming the core of the Cardiff–Cowbridge anticline. In these widely scattered occurrences they display much variation in facies and provide some hints of contemporary palaeogeography.

About Tarannon and Talerddig in Montgomeryshire a lower group, the Nant-ysgollon Shales of the *murchisoni* and *riccartonensis* zones, 400 ft thick, is followed by an upper group, the Fynyddog Grits, alternating shales and coarse sandstones to a thickness of 1500 ft. The sandy beds diminish in importance southwards, but silty flagstones interbedded with striped shales and mudstones reach thicknesses of nearly 3000 ft about Builth, and 2500 ft towards Llandeilo, beyond which however they thin to about 900 ft. Slumped and convoluted beds recur at a number of horizons, and many of the beds are greywackes. The most common fossils in the outcrops between Llanidloes and Builth are graptolites, of which species of *Cyrtograptus*, with *Monograptus flemingii* and *M. riccartonensis*, are notable, and the lithological and faunal aspect of the sequence is geosynclinal or basinal. Towards Llandeilo, however, while the rock suite is not greatly different and some beds yield graptolites including *Monograptus priodon* and *M. vomerinus*, there are intercalated impure lime-stones and calcareous mudstones, and shelly fossils become increasingly common and include species characteristic of the 'shelf' facies of the Welsh borderland—*Resserella* [*Dalmanella*] *elegantula, Dicoelosia* [*Bilobites*] *biloba, Skenidioides lewisii, Plectodonta transversalis, Eospirifer radiatus, Cyrtia exporrecta, Atrypa reticularis, Meristina obtusa, Dalmanites caudatus, Encrinurus punctatus, Calymene blumenbachii, Phacops stokesii.*

Signs of uplift with contemporary warping, and of a migratory shore, are seen in the development of an unconformity with overstep at the base of the upper Wenlock strata (of olive green mudstones and sandstones), about 700 ft of lower Wenlock (of laminated striped grey mudstones) being reduced to a feather-edge, apparently by overlap, just east of Llandeilo. Immediately to the south-west there is a transgression of upper Wenlock beds on to Llandeilo Flags and Llanvirn shales—a transgression that takes on the magnitude of the sub-Wenlock unconformity, and is a further sign of the restlessness of the terrain to the east and south-east of what is now the Towy axis.

The Wenlock Series in shelly facies is well developed in the Pembrokeshire outcrops. At Marloes and Wooltack, on the mainland east of Skomer, the thickness is reduced to 400–500 ft. The rocks are grey shales, mudstones, and calcareous sandstones forming the upper part of the Coralliferous Series of De la Beche. They are abundantly fossiliferous, with many brachiopods (*Atrypa reticularis, Eospirifer radiatus, Delthyris elevata, Strophonella euglypha, Leptaena rhomboidalis*), trilobites (*Calymene, Bumastus, Acaste, Dalmanites*), and corals (*Favosites*). The beds are particularly calcareous towards the base, where an impure limestone lies at about the horizon of the Woolhope Limestone of the Welsh borders. The whole aspect of the rock suite is neritic in shelf facies, not geosynclinal; and while there seems to be conformable transition from underlying Llandovery in the coastal outcrops, the Wenlock sediments overstep on to what must have been an island or promontory of Skomer volcanic rocks (of Arenig age) at Marloes Mill, where a contemporary shore may be identified. A few miles farther south-east, at Freshwater East, residual Wenlock beds are reduced to less than 40 ft of highly fossiliferous calcareous mudstones and sandstones. They overstep with strong unconformity on to a contemporary submergent coast of Llanvirn shales.

In the Rumney inlier at Cardiff the Wenlock rocks match their equivalents in Pembrokeshire in being calcareous mudstones, sandstones, and flaggy siltstones with a rich shelly fauna. A prominent sandy bed towards the top of the series, the Rumney Grit, is strongly cross-bedded and ripple-marked, and contains seams of carbonaceous debris: it suggests deposition at no great distance from a contemporary shore.

Ludlow Series

The Ludlow rocks appear to follow the Wenlock everywhere without break, and locally are not easily distinguished from them. They form wide outcrops flanking the Wenlock around the Builth Ordovician inlier; but south-west of Builth the outcrop becomes progressively narrower because of overstep by the Old Red Sandstone, and between Llandeilo and Carmarthen it is finally extinguished. Beds of the series emerge with the Wenlock in the small outcrops of south-west Pembrokeshire, and in the Rumney inliers.

In the Builth district, where the work of Straw has been particularly fruitful, the full thickness of the series may approach 5000 ft, in a 'basinal' sequence that shows marked variation in rock-type and inferred conditions of sedimentation. The lowest beds comprise about 750 ft of graptolitic shales, flags, and mudstones that follow the Wenlock strata without notable change in a 'geosynclinal' lithology. They are distinguished by the incoming of graptolites of the type of *Monograptus colonus* and by the disappearance of *Cyrtograptus*. They are followed by some 1500 ft of conspicuously slumped shelly siltstones, calcareous rocks in which convolutions many feet in amplitude give the impression of tectonic folds: the direction of slumping usually runs northwards or north-westwards from shallows in the southern part of South Wales. Slumps are rare in higher beds, and the upper part of the sequence, in a thickness of about 2500 ft, is of laminated siltstones becoming increasingly shelly and neritic, with an abundance of brachiopods (*Camarotoechia nucula, Sphaerirhynchia* [*Wilsonia*] *wilsoni, Howellella* [*Delthyris*] *elegans, Dayia navicula, Whitfieldella* spp., *Salopina* [*Dalmanella*] *lunata, Protochonetes ludloviensis* [*Chonetes striatellus*], *Chonetoidea grayi*), bivalves (*Fuchsella* [*Orthonota*] *amygdalina, Cardiola interrupta, Modiolopsis complanata*), gastropods (*Loxonema* [*Holopella*] spp., *Platyschisma helicites*), and ostracods (*Neobeyrichia* [*Beyrichia*] *lauensis*), together with such trilobites as *Dalmanites, Calymene,* and *Encrinurus*.

South-westwards from Builth lateral changes include a diminution in thickness of the series to about 2000 ft by systematic attenuation of each member, a disappearance of the graptolitic shales at the base and of slumped beds in the overlying siltstones, an increase in the number of impure shelly limestones and decalcified rottenstones, and an incoming near Llandeilo of coarse cross-bedded and ripple-marked delta sediments both in the lower part of the sequence (the Lletty Grit) and in the middle (the Trichrûg Beds, which reach 600 ft in a closely defined lenticle). The grits show signs of a proximate origin of their detritus in a coastal region to the south (now hidden by the blanket of Old Red Sandstone), the Trichrûg member being underlain by the highly local *Grammysia* Beds, purple mudstones and sandstones with a rich molluscan fauna including the bivalve *Grammysia triangulata*. The succession, as described by Potter and Price, shows the kind of variation characteristic of a near-shore mixed-facies environment; and repeated suggestions of non-sequence particularly in the south-westernmost outcrops imply a pulsatory subsidence of the depositional floor and interludes of slight uplift. (*See* Fig. 16.)

In south-west Pembrokeshire Ludlow rocks in shelf facies are mainly calcareous sandstones and mudstones like the uppermost Wenlock, and the base is difficult to define. At Wooltack and Marloes they are about 2000 ft thick, relatively unfossiliferous. At Freshwater West they are reduced to 250 ft, but are more calcareous—green sandstones and shelly mudstones containing thin bands of impure limestone with characteristic brachiopods, molluscs, ostracods, and trilobites (including *Homalonotus johannis* and *Acaste downingiae*). In places they overlap Wenlock mudstones to rest on Llanvirn shales, and are themselves almost completely overstepped by the Old Red Sandstone. At Freshwater East, as sign of the shallow waters of a near-shore environment, they rest with conglomeratic base discordantly on Wenlock beds.

In the Rumney inlier the series, though not well exposed, is represented by about 300 ft of flaggy sandstones, mudstones, and shales. The highest

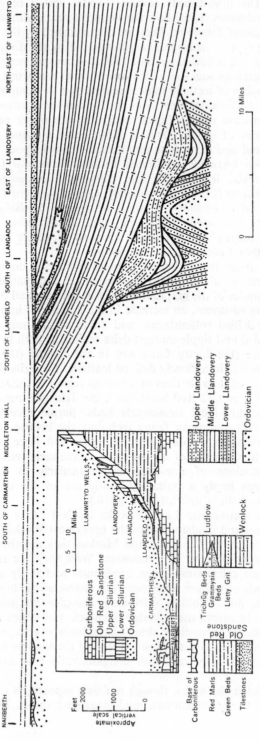

FIG. 16. *Section of the Salopian rocks along the eastern flank of the Towy anticline, to show their unconformable relations with the Llandovery rocks, and their disappearance westwards into Pembrokeshire through overstep by the Old Red Sandstone.*

(In part after Straw and his colleagues.)

bed is a brown grit, full of fish-remains including spines of *Onchus tenuis-triatus*, that suggests, but may be older than, the Ludlow Bone Bed at the base of the Old Red Sandstone.

References

Ordovician and Silurian

ADAMS, T. D. 1962. The geology of the Dinas–Cwm Rheidol hydroelectric tunnel. *Geol. Mag.*, **100**, 317–38.

ANDERSON, J. G. C. 1963. The geology of the Rheidol hydroelectric project. *Proc. S. Wales Inst. Eng.*, **78**, 35–47.

ANDREW, G. 1925. The Llandovery rocks of Garth, Breconshire. *Quart. J. Geol. Soc.*, **81**, 389–406.

—— and JONES, O. T. 1925. The relations between the Llandovery rocks of Llandovery and those of Garth. *Quart. J. Geol. Soc.*, **81**, 407–16.

BAILEY, R. J. 1964. A Ludlovian facies boundary in south-central Wales. *Liv. & Manch. Geol. J.*, **4**, 1–19.

—— 1969. Ludlovian sedimentation in south-central Wales. *In* A. Wood (editor): *Pre-Cambrian and Lower Palaeozoic rocks of Wales*, 283–304. Cardiff.

BASSETT, D. A. 1955. The Silurian rocks of the Talerddig district, Montgomeryshire. *Quart. J. Geol. Soc.*, **111**, 239–64.

—— 1963. The Welsh Palaeozoic geosyncline: a review of recent work on stratigraphy and sedimentation. *In* M. R. W. Johnson and F. H. Stewart (editors): *The British Caledonides*, 35–69. Edinburgh.

CANTRILL, T. C. and THOMAS, H. H. 1906. The igneous and associated sedimentary rocks of Llangynog. *Quart. J. Geol. Soc.*, **62**, 223–52.

COX, A. H. 1915. The geology of the district between Abereiddy and Abercastle. *Quart. J. Geol. Soc.*, **71**, 273–342.

CROSFIELD, M. C. and SKEAT, E. G. 1896. On the geology of the neighbourhood of Carmarthen. *Quart. J. Geol. Soc.*, **52**, 523–63.

CUMMINS, W. A. 1959. The Lower Ludlow grits in Wales. *Liv. & Manch. Geol. J.*, **2**, 168–79.

—— 1969. Patterns of sedimentation in the Silurian rocks of Wales. *In* A. Wood (editor): *Pre-Cambrian and Lower Palaeozoic rocks of Wales*, 219–37. Cardiff.

DAVIES, K. A. 1926. The geology of the country between Drygarn and Abergwesyn, Breconshire. *Quart. J. Geol. Soc.*, **82**, 436–64.

—— 1928. Notes on the geology of the southern portion of Central Wales. *Proc. Geol. Assoc.*, **39**, 157–60.

—— 1928. The geology of the country between Rhayader (Radnorshire) and Abergwesyn (Breconshire). *Proc. Geol. Assoc.*, **39**, 160–8.

—— 1933. The geology of the country between Abergwesyn (Breconshire) and Pumpsaint (Carmarthenshire). *Quart. J. Geol. Soc.*, **89**, 172–201.

—— and PLATT, J. I. 1933. The conglomerates and grits of the Bala and Valentian rocks. *Quart. J. Geol. Soc.*, **89**, 202–20.

DREW, H. and SLATER, I. L. 1910. Notes on the geology of the district around Llansawel, Carmarthenshire. *Quart. J. Geol. Soc.*, **66**, 402–19.

ELLES, G. L. 1900. The zonal classification of the Wenlock Shales of the Welsh borderland. *Quart. J. Geol. Soc.*, **56**, 370–414.

—— 1939. The stratigraphy and faunal succession in the Ordovician rocks of the Builth–Llandrindod inlier, Radnorshire. *Quart. J. Geol. Soc.*, **95**, 338–445.

ELSDEN, J. V. 1905. The igneous rocks occurring between St. David's Head and Strumble Head. *Quart. J. Geol. Soc.*, **61**, 579–607.

EVANS, D. C. 1906. The Ordovician rocks of western Carmarthenshire. *Quart. J. Geol. Soc.*, **62**, 605–43.

EVANS, W. D. 1945. The geology of the Prescelly Hills, north Pembrokeshire. *Quart. J. Geol. Soc.*, **101**, 89–110.

—— 1948. The Cambrian–Ordovician junction, Whitesand Bay, Pembrokeshire. *Geol. Mag.*, **85**, 110–12.

HENDRIKS, E. M. L. 1926. The Bala–Silurian succession in the Llangranog district (south Cardiganshire). *Geol. Mag.*, **63**, 121–34.

HICKS, H. 1873. On the Tremadoc rocks in the neighbourhood of St. David's. *Quart. J. Geol. Soc.*, **29**, 39–52.

—— 1875. On the succession of ancient rocks in the vicinity of St. David's. *Quart. J. Geol. Soc.*, **31**, 167–95.

HOLLAND, C. H. 1959. The Ludlovian and Downtonian rocks of the Knighton district, Radnorshire. *Quart. J. Geol. Soc.*, **114**, 449–82.

—— 1969. The Welsh Silurian geosyncline in its regional context. *In* A. Wood (editor): *Pre-Cambrian and Lower Palaeozoic rocks of Wales*, 203–17. Cardiff.

—— and LAWSON, J. D. 1963. Facies patterns in the Ludlovian of Wales and the Welsh borderland. *Liv. & Manch. Geol. J.*, **3**, 269–88.

JONES, O. T. 1909. The Hartfell–Valentian succession in the district around Plynlimon and Pont Erwyd, north Cardiganshire. *Quart. J. Geol. Soc.*, **65**, 463–537.

—— 1921. The Valentian series. *Quart. J. Geol. Soc.*, **77**, 244–74.

—— 1925. The geology of the Llandovery district: Part I—The southern area. *Quart. J. Geol. Soc.*, **81**, 344–88.

—— 1935. The Lower Palaeozoic rocks of Britain. *Rep. 16th Internat. Geol. Congr.* (Washington), **1**, 463–84.

—— 1938. On the evolution of a geosyncline. *Quart. J. Geol. Soc.*, **94**, lx–cx.

—— 1947. The geology of the Silurian rocks west and south of the Carneddau range, Radnorshire. *Quart. J. Geol. Soc.*, **103**, 1–36.

—— 1949. The geology of the Llandovery district: Part II—The northern area. *Quart. J. Geol. Soc.*, **105**, 43–70.

—— and PUGH, W. J. 1915. The geology of the district around Machynlleth and the Llyfnant valley. *Quart. J. Geol. Soc.*, **81**, 343–85.

—— —— 1935. The geology of the districts around Machynlleth and Aberystwyth. *Proc. Geol. Assoc.*, **46**, 247–300.

—— —— 1941. The Ordovician rocks of the Builth district. *Geol. Mag.*, **78**, 185–91.

—— —— 1946. The complex intrusion of Welfield, near Builth Wells, Radnorshire. *Quart. J. Geol. Soc.*, **102**, 157–88.

—— —— 1948. A multi-layered dolerite complex of laccolithic form, near Llandrindod Wells, Radnorshire. *Quart. J. Geol. Soc.*, **104**, 43–70.

—— —— 1948. The form and distribution of dolerite masses in the Builth–Llandrindod inlier, Radnorshire. *Quart. J. Geol. Soc.*, **104**, 71–98.

—— —— 1949. An early Ordovician shoreline in Radnorshire, near Builth Wells. *Quart. J. Geol. Soc.*, **105**, 65–99.

JONES, W. D. V. 1945. The Valentian succession around Llanidloes, Montgomeryshire. *Quart. J. Geol. Soc.*, **100**, 309–32.

KELLING, G. 1964. The turbidite concept in Britain. *In* A. H. Bouma and A. Brouwer: *Turbidites*, 75–92. Amsterdam.

KELLING, G. and WOOLLANDS, M. A. 1969. The stratigraphy and sedimentation of the Llandoverian rocks of the Rhayader district. *In* A. Wood (editor): *Pre-Cambrian and Lower Palaeozoic rocks of Wales*, 255–92. Cardiff.

LAPWORTH, H. 1900. The Silurian sequence of Rhayader. *Quart. J. Geol. Soc.*, **56**, 67–137.

MARR, J. E. and ROBERTS, T. 1885. The Lower Palaeozoic rocks of the neighbourhood of Haverfordwest. *Quart. J. Geol. Soc.*, **41**, 476–91.

MURCHISON, R. I. 1839. *The Silurian system*. London.

POTTER, J. F. and PRICE, J. H. 1966. Comparative sections through rocks of Ludlovian–Downtonian age in the Llandovery and Llandeilo districts. *Proc. Geol. Assoc.*, **76**, 379–402.

PRINGLE, J. 1930. The geology of Ramsey Island, Pembrokeshire. *Proc. Geol. Assoc.*, **41**, 1–31.

RAST, N. 1969. The relationship between Ordovician structure and vulcanicity in Wales. *In* A. Wood (editor): *Pre-Cambrian and Lower Palaeozoic rocks of Wales*, 305–35. Cardiff.

RICH, J. L. 1950. Floor markings, groovings, interstratal crumplings as criteria for recognition of slope deposits, with illustrations from Silurian rocks of Wales. *Bull. Amer. Assoc. Petrol. Geol.*, **34**, 717–41.

ROACH, R. A. 1969. The composite nature of the St. David's Head and Carn Llidi intrusions of north Pembrokeshire. *In* A. Wood (editor): *Pre-Cambrian and Lower Palaeozoic rocks of Wales*, 409–33. Cardiff.

ROBERTS, R. O. 1927. The igneous and associated Ordovician rocks of Baxter's Bank, Radnorshire. *Geol. Mag.*, **64**, 289–98.

—— 1929. The geology of the district around Abbey-Cwm-hir, Radnorshire. *Quart. J. Geol. Soc.*, **85**, 651–76.

SKEVINGTON, D. 1969. The classification of the Ordovician system in Wales. *In* A. Wood (editor): *Pre-Cambrian and Lower Palaeozoic rocks of Wales*, 161–79. Cardiff.

SMITH, A. J. and LONG, G. H. 1969. The Upper Llandovery sediments of Wales and the Welsh borderlands. In A. Wood (editor): *Pre-Cambrian and Lower Palaeozoic rocks of Wales*, 239–53. Cardiff.

STAMP, L. D. and WOOLDRIDGE, S. W. 1923. The igneous and associated rocks of Llanwrtyd, Brecon. *Quart. J. Geol. Soc.*, **79**, 16–46.

STRAW, S. H. 1930. The Siluro–Devonian boundary in south central Wales. *J. Manch. Geol. Assoc.*, **1**, 79–102.

—— 1937. The higher Ludlovian rocks of the Builth district. *Quart. J. Geol. Soc.*, **93**, 406–56.

—— 1952. The Silurian succession at Cwm Graig Ddu (Breconshire). *Liv. & Manch. Geol. J.*, **1**, 208–19.

THOMAS, G. E. and THOMAS, T. M. 1957. The volcanic rocks of the area between Fishguard and Strumble Head, Pembrokeshire. *Quart. J. Geol. Soc.*, **112**, 291–311.

THOMAS, H. H. 1911. The Skomer volcanic series, Pembrokeshire. *Quart. J. Geol. Soc.*, **67**, 175–214.

—— and COX, A. H. 1924. The volcanic series of Trefgarn, Roch, and Ambleston. *Quart. J. Geol. Soc.*, **80**, 520–48.

WILLIAMS, A. 1953. The geology of the Llandeilo district, Carmarthenshire. *Quart. J. Geol. Soc.*, **108**, 177–208.

—— 1969. Ordovician faunal provinces with reference to brachiopod distribution. *In* A. Wood (editor): *Pre-Cambrian and Lower Palaeozoic rocks of Wales*, 117–54. Cardiff.

WOOD, A. and SMITH, A. J. 1959. The sedimentation and sedimentary history of the Aberystwyth Grits (Upper Llandoverian). *Quart. J. Geol. Soc.*, **114**, 163–95.

WOOD, E. M. R. 1900. The Lower Ludlow formation and its graptolite-fauna. *Quart. J. Geol. Soc.*, **56**, 415–92.

—— 1906. On the Tarannon Series of Tarannon. *Quart. J. Geol. Soc.*, **62**, 644–701.

ZIEGLER, A. M. 1965. Silurian marine communities and their environmental significance. *Nature*, **207**, 270–2.

—— COCKS, L. R. M. and MCKERROW, W. S. 1968. The Llandovery transgression in the Welsh borderland. *Palaeontology*, **11**, 736–83.

6. Old Red Sandstone

Before the close of Silurian times much of Britain was affected by strong earth-movements that caused uplift and sharp folding, and a final destruction of the Lower Palaeozoic geosyncline. The movements in Wales, which were the culmination of a crustal restlessness recurrent since Cambrian times, brought into being a tract of upland, 'St. George's Land', that occupied much of the central and northern parts of the country and probably extended westwards towards Ireland and eastwards into the English midlands. With the uplift the sea retreated southwards and became restricted to the region running through the present Devon and Cornwall, where fossiliferous shallow-water marine shales and limestones continued to be deposited as the Devonian rocks. Between St. George's Land and the Devonian sea was a tract, perhaps a gulf or embayment, in which terrigenous sediments, composed of detritus eroded from the mountainous hinterland to the north, were laid down under fluviatile and deltaic conditions. As 'continental' non-marine and mostly unfossiliferous marls and sandstones, usually stained by hematite, they are the rocks of the Old Red Sandstone—a formation of Devonian age, in transition southwards into the rocks of the Devonian sea. The basin of deposition of the Old Red Sandstone extended into southern England, and is called the Anglo-Welsh cuvette.

Sedimentation was virtually unbroken throughout Devonian times in the Devon sea, but north of the Bristol Channel earth-movement including some folding was renewed in mid-Devonian times when South Wales was uplifted into a zone of erosion. In consequence, the rock-sequence was interrupted and now falls into two clearly distinct parts; Lower Old Red Sandstone is everywhere followed with strong unconformity by Upper, and Middle Old Red Sandstone is unknown.

The fossils of the Old Red Sandstone include the first vascular plants found in South Wales, and (in any number) the first fishes. In the Lower Old Red Sandstone jawless ostracoderms, in some ways comparable with living lampreys, reached their acme in a variety of armoured types. In the Upper Old Red Sandstone true bony fishes are found in diagnostic genera.

Silurian Ludlow beds are followed by the Old Red Sandstone without notable break in the Welsh borderland, where a group of transition rocks, ranging upwards from marine to continental, does not readily fall into either system. Although disagreement on where the junction between the two systems is best placed has persisted since Murchison's day, the readily identifiable Ludlow Bone Bed has commonly been used as an acceptable field marker of the base of the Old Red Sandstone; and it has the further convenience in South Wales of being the horizon at which a strong unconformity is developed westwards at the base of the Old Red Sandstone.

Lower Old Red Sandstone

Downton Series

Above the horizon of the Ludlow Bone Bed the Tilestones, broadly equivalent to the Downton Castle Sandstone and the Temeside Shales of Shropshire, consist of about 150 ft of grey, green, and yellow micaceous flaggy sandstones with some grits and rottenstones. As 'grey Downtonian' they are less thoroughly continental than the overlying rocks, and although they are not open-water marine sediments like the underlying bluish grey calcareous Ludlow Shales they contain a mixed if restricted fauna of brachiopods, bivalves, gastropods, and ostracods, including *Lingula cornea, Salopina* [*Dalmanella*] cf. *lunata, Camarotoechia nucula, Grammysia extrasulcata, Modiolopsis complanata, Platyschisma helicites,* 'eurypterids', and *Neobeyrichia wilckensiana.* Fish-remains are relatively rare, but form a bone-bed in the Rumney section at Cardiff. At Capel Horeb near Llangadock *Thallomia breconensis,* an early vascular plant showing well-preserved stomata, occurs in an interbedded lens of shale.

Overlying rocks are dominantly red and fine-grained, and constitute the Downtonian Red Marls, over 2000 ft thick. In detail they are alternating mudstones, siltstones, and fine sandstones, many of the beds displaying current rippling and fine cross-bedding. Despite their red colour they contain occasional layers of fossils including brachiopods and molluscs, and many beds were reworked by scavengers and burrowers. Their general characteristics suggest they were deposited in the brackish waters of wide delta flats not far above sea-level, where waves and currents moulded the beds, and where periodic incursions of the sea allowed the formation of marine or near-marine intercalations. Towards or at the top of the group, and continuing for about 100 ft into the overlying Ditton Series, impersistent bands of nodular and conglomeratic calcareous 'cornstones', probably the product of desiccation, are rich in fish remains, and as the '*Psammosteus*' Limestones (with *Traquairaspis* [*Psammosteus anglicus*]) form a widespread marker of great use in mapping.

Westwards there was, however, a pre-Devonian up-tilting of the flanks of St. George's Land, and transition from Ludlow into Downtonian sediments is lost. There is increasing intensity of overstep and a final extinction of the whole Silurian sequence; and from Carmarthen into Pembrokeshire the Old Red Sandstone rests in places on Ordovician rocks down to Arenig. The basal Downtonian strata, the Tilestones, at the same time either pass laterally into, or more probably are overlapped by, the Green Beds, micaceous sandstones and mudstones sometimes with a pebbly base. (*See* Fig. 16.)

Many kinds of ostracoderms are found in the Downtonian rocks, of which cephalaspids (*Hemicyclaspis, Traquairaspis*) are dominant types. Whole skeletons are rare, the disarticulation of the fossils, and the usual occurrence only of isolated plates and shields, being signs of the lively waters in which the sediments accumulated.

Ditton Series

The general lithology of the Downtonian rocks is continued upwards in the Dittonian Red Marls, about 1500 ft thick. In detail they are, like the beds beneath, a mixed group of mudstones and fine sandstones, but they are almost without marine intercalations and they contain a number of cornstone layers: they appear to have accumulated mainly as delta fans and flood-plain sediments. They display scour-and-fill structures, and washouts are not uncommon. Bedding-planes are covered with rain-pits or are suncracked. Conglomeratic layers lie immediately above slightly eroded surfaces. Burrows and invertebrate fossils are rare, but drifted plants form carbonaceous films.

The Dittonian ostracoderm fauna is dominated, and is distinguished from the Downtonian, by species of pteraspids (*Cymripteraspis leachi, Belgicaspis crouchi, Rhinopteraspis cornubica*) long-snouted arrivals in the Welsh province found in association with cephalaspids. Their rapid evolution and the readily recognizable shapes of the skeletal plates of the different species make them well suited for use as zonal forms.

Brecon Series

Along the escarpment face of the Black Mountains, the Brecon Beacons, and the Fans the Red Marls are followed by a group of sandstones, reaching 2000 to 3000 ft in thickness, that (like the Marls) persist with little change for many miles along their outcrop. They range from siltstones and flaggy mudstones to grits and some conglomerates, and they display a rhythmic alternation of coarse and fine beds through many hundreds of units that gives them a characteristic appearance in the scarp face (*see* Pl. XIIA). They fall into two distinct formations, the Senni Beds below, with dark green chloritic layers interbedded with red, and the Brownstones above, uniformly dark red and purple. The differences in colour may be due to a richer organic content in the Senni Beds, which have yielded not only the last of the pteraspids but also a flora of some variety including *Psilophyton, Dawsonites*, and *Gosslingia*—whereas the Brownstones have as yet yielded no fossils. In places the sequence above the Red Marls is uniformly brown or purple, and a lithologically recognizable group of Senni Beds cannot be distinguished from the Brownstones, probably because of lateral passage of green beds into brown.

The presence of pteraspids (*Rhinopteraspis cornubica, Penygaspis dixoni*) in the Senni Beds places them in the Lower Old Red Sandstone. The Brownstones, in their transitional conformity and diachronous merging with the Senni Beds, are not significantly different in age and also are referable to the Lower Old Red Sandstone.

The steady maintenance of a uniform sequence in the Lower Old Sandstone from Shropshire across the wide Breconshire outcrops into Carmarthenshire breaks down farther west in Pembrokeshire where two formations highly localized in occurrence are intercalated in the Breconian sequence—the Ridgeway Conglomerate and the Cosheston Beds. The Ridgeway Conglomerate, which attains a thickness of 1200 ft, is found only south of Milford Haven, where it immediately overlies the Red Marls and underlies transgressive Upper Old Red Sandstone. While a regional palaeogeography suggests accumulation of the Old Red Sandstone on the southern flanks of St. George's Land, some quartzite pebbles in the Conglomerate are foreign, wholly unlike any that could have been derived from source rocks in mid-Wales, and compare with rock-types now known only in the Grès de May in France. Some of the pebbles are large and imperfectly rounded: they appear not to have travelled any great distance, and they hint at a massif of ancient rock not far away to the south. Similar sediments near Cardiff, the Llanishen Conglomerate, at about the same Breconian horizon, are a further pointer to a southerly source of origin.

The Cosheston Beds of Pembrokeshire, found mainly north of Milford Haven, are less exotic. Their lower members, which directly follow the Red Marls, are micaceous sandstones and marls very like the Senni Beds; but higher in the sequence, where Brownstones normally occur, the rocks are breccias and conglomerates composed in great part of pebbles and fragments of igneous rock—although igneous rocks of Devonian age are nowhere known in South Wales. The group reaches a thickness of about 10 000 ft, and is terminated only by overstep of the Upper Old Red Sandstone. (*See* Fig. 17.)

Upper Old Red Sandstone and Upper Devonian

No representatives of the Middle Old Red Sandstone being known to occur in South Wales, the Upper Old Red Sandstone rests with unconformity upon the beds beneath. It consists of a varied group of sandstones, grits, and conglomerates variable in development. In Pembrokeshire the beds comprise the Skrinkle Sandstones; on the flanks of the South Wales coalfield they include the Quartz Conglomerates, the Plateau Beds, and the Grey Grits. In Pembrokeshire particular interest attaches to the occurrence of marine intercalations that link the upper beds of the Skrinkle Sandstones with the Upper Devonian rocks of Devon.

Skrinkle Sandstones

The greater part of the Skrinkle Sandstones is in the 'continental' facies of the Old Red Sandstone and comprises a mixed suite of sandstones, conglomerates, breccias and red marls. Many of the subangular fragments in the breccias are of igneous rocks, ashy, rhyolitic, and granitic, like those of the Pre-Cambrian and Ordovician formations of north Pembrokeshire. They imply exposure of source rocks at no great distance, and therefore absence of Lower Old Red Sandstone, in an intensification of the mid-Devonian unconformity northwards. Scales of *Holoptychius*, a fringe-finned crossopterygian ganoid fish, and some bony armour have been found near the base and prove a reference of the rocks to the Upper Old Red Sandstone. The beds are over 1000 ft thick at Freshwater West but decrease to about 400 ft near West Angle.

The uppermost beds of the Skrinkle Sandstones show alternations of 'continental' red beds and grey and green marine shales, mudstones, siltstones, and highly fossiliferous impure limestones. As was first shown by Salter more than a hundred years ago, the fossils closely resemble those of the Marwood Beds of Devon, of Upper Devonian age: they include crinoids, bryozoans, brachiopods (*Cyrtospirifer verneuili*, rhynchonelloids), bivalves

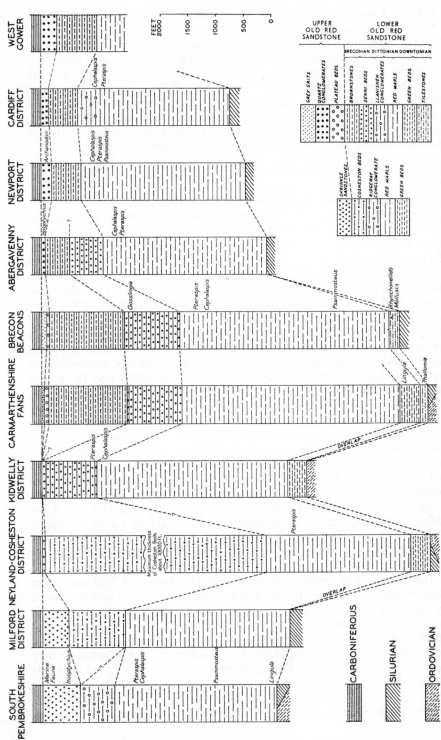

Fig. 17. *Comparative columns showing variation in thickness and sequence of the Old Red Sandstone*

(scallops, *Cucullaea*, *Ptychopteria damnoniensis*, modiolids, sanguinolitids), cephalopods, gastropods, ostracods, and fishes. Plant remains are common, but not well preserved: they include woody stems (*Bothrodendron?*) and fern-like leaflets (*Sphenopteris?*): at two horizons they are sufficiently abundant to form impure coal lenticles. The beds merge without a break into the Carboniferous Lower Limestone Shales. They reach about 160 ft in southernmost outcrops but thin and become increasingly 'continental' northwards. Neither they nor the underlying beds of the Skrinkle group are found north of the line of the Ritec fault, a structure that in incipient growth may have defined their depositional limits.

Quartz Conglomerates, Plateau Beds, and Grey Grits

Farther east, in the outcrops surrounding the South Wales coalfield, the sediments of the Upper Old Red Sandstone fall into three depositional units. Over most of the ground they are represented by the Quartz Conglomerate Group, red and brown sandstones, quartzites, and coarse conglomerates whose identifiable constituents, of metamorphic and igneous rocks, vein quartz, and greywackes, lie in a sandy matrix rich in detrital feldspars and suggest derivation from sources to the north, including rock-suites like those of the Mona Complex in Anglesey and of the Lower Palaeozoic rocks of mid-Wales and North Wales. In Allen's interpretation, their sedimentological features characterize deposits formed by fast-stream transport along several major channels radiating southwards. They are relatively unfossiliferous but have yielded scales of *Holoptychius* in the Usk valley and the mussel *Archanodon Jukesi* near Newport. Drifted plant debris is not uncommon, sometimes concentrated in pockets.

In Carmarthenshire and Breconshire, where they form a table-top cap to the Brecon Beacons (*see* cover photograph), the Plateau Beds overlie the Brownstones in visibly discordant overstep. They may be in part the lateral equivalents of the Quartz Conglomerate Group, or (as Allen has suggested) they may be slightly older. They are tough red quartzites, often conglomeratic, with pebble and mineral constituents like those of the Quartz Conglomerates. A particular feature in their outcrops flanking the Tawe valley is the occurrence of a marine bed containing fossils, including *Cyrtospirifer verneuili*, like some found in the Skrinkle Sandstones, that prove not only an Upper Devonian age but also a brief incursion of the Skrinkle sea.

The Plateau Beds reach a thickness of more than 100 ft at maximum development, but thin westwards and are finally overlapped or overstepped by the Grey Grits, a group sharply contrasting with the beds beneath in their colour and in being unfossiliferous clean-washed sandstones and pebble beds, stream-channel deposits of fluviatile origin. They form lenticles rarely more than 20 ft thick but locally expanding to 200 ft. Eastwards they perhaps pass laterally into the upper part of the Quartz Conglomerates; westwards they are conformably overlapped by the Lower Limestone Shales, which transgress on to the Brownstones and Senni Beds of the Lower Old Red Sandstone between Llandebie and Kidwelly.

References

ALLEN, J. R. L. 1962. Intraformational conglomerates and scoured surfaces in the Lower Old Red Sandstone of the Anglo-Welsh cuvette. *Liv. & Manch. Geol. J.*, **3**, 1–20.

—— 1963. Depositional features of Dittonian rocks: Pembrokeshire compared with the Welsh borderland. *Geol. Mag.*, **100**, 385–400.

—— 1964. Studies in fluviatile sedimentation: six cyclothems from the Lower Old Red Sandstone, Anglo-Welsh basin. *Sedimentology*, **3**, 163–98.

—— 1964. The pre-Pickwell Down age of the Plateau Beds (Upper Devonian) in South Wales. *Nature*, **204**, 364–6.

—— 1964. Primary current direction in the Lower Old Red Sandstone (Devonian), Anglo-Welsh basin. *Sedimentology*, **3**, 89–108.

—— 1965. Upper Old Red Sandstone (Farlovian) paleogeography in South Wales and the Welsh borderland. *J. Sed. Pet.*, **35**, 167–95.

—— and TARLO, L. B. 1963. The Downtonian and Dittonian facies of the Welsh borderland. *Geol. Mag.*, **100**, 129–55.

CROFT, W. N. 1953. Breconian: a stage name of the Old Red Sandstone. *Geol. Mag.*, **90**, 429–33.

—— and LANG, W. H. 1942. The Lower Devonian flora of the Senni Beds of Monmouthshire and Breconshire. *Phil. Trans. Roy. Soc.*, (B), **231**, 131–63.

DIXON, E. E. L. 1933. Some recent stratigraphical work in its bearing on south Pembrokeshire problems. *Proc. Geol. Assoc.*, **44**, 217–25.

HEARD, A. 1927. On Old Red Sandstone plants showing structure, from Brecon, South Wales. *Quart. J. Geol. Soc.*, **83**, 195–209.

—— 1939. Further notes on Lower Devonian plants in South Wales. *Quart. J. Geol. Soc.*, **95**, 223–9.

—— and DAVIES, R. 1924. The Old Red Sandstone of the Cardiff district. *Quart. J. Geol. Soc.*, **75**, 489–519.

—— and JONES, J. F. 1931. A new plant (*Thallomia*) showing structure from the Downtonian rocks of Llandovery, Carmarthenshire. *Quart. J. Geol. Soc.*, **87**, 551–62.

KING, W. W. 1934. The Downtonian and Dittonian strata of Great Britain and north-western Europe. *Quart. J. Geol. Soc.*, **90**, 526–70.

LANG, W. H. 1937. On the plant-remains from the Downtonian of England and Wales. *Phil. Trans. Roy. Soc.*, (B), **227**, 254–91.

POTTER, J. F. 1967. Deformed micaceous deposits in the Downtonian of the Llandeilo region, South Wales, *Proc. Geol. Assoc.* **78**, 277–88.

STRAW, S. H. 1930. The Siluro-Devonian boundary in south-central Wales. *J. Manch. Geol. Assoc.*, **1**, 79–102.

WHITE, E. I. 1938. New pteraspids from South Wales. *Quart. J. Geol. Soc.*, **94**, 85–115.

—— 1946. The genus *Phialaspis* and the '*Psammosteus* Limestones'. *Quart. J. Geol. Soc.*, **101**, 207–42.

—— 1950. The vertebrate faunas of the Lower Old Red Sandstone of the Welsh borders. *Bull. Brit. Mus. (Nat. Hist.)*, **1**, 51–67.

WILLIAMS, D. M. 1926. Notes on the relation of the Upper and Lower Old Red Sandstones of Gower. *Geol. Mag.*, **63**, 219–23.

7. Carboniferous Limestone

The marine Devonian intercalations in the Upper Old Red Sandstone of Pembrokeshire marked the beginnings of a general subsidence of the southern flanks of St. George's Land, when the sea, mainly confined in Devonian times to the area south of the (present) Bristol Channel, advanced northwards and flooded the freshwater cuvette of the Old Red Sandstone. The physique of the region was not greatly modified by earth-movement as the Devonian period passed into the Carboniferous, but the environment of sedimentation was radically changed by the subsidence, and the rocks of the transgressive sea differ markedly from the 'continental' fluviatile and deltaic red sandstones, conglomerates, and marls of the Upper Old Red Sandstone in consisting of grey richly fossiliferous calcareous shales and massive limestones, reaching in places a thickness of over 4000 ft. They are called the Carboniferous Limestone or Avonian or Dinantian Series.[1]

The gentle transgression of the Carboniferous sea is clearly shown in south Pembrokeshire, where the fossils in the junction beds linking the Skrinkle Sandstones and the Lower Limestone Shales have both Devonian and Avonian affinities. At all localities where they are seen in contact the Avonian rocks rest conformably on the Upper Old Red Sandstone. At the same time, being the deposits of a transgressive sea, they overlap the Upper Old Red Sandstone in places and extend northwards unconformably on to older strata. Thus the Lower Limestone Shales progressively overstep the Brownstones west from Llandebie, and come to rest on the Senni Beds at Kidwelly; and they overstep on to the Red Marls along the north crop in Pembrokeshire, with a development of a coarse basal conglomerate near Haverfordwest.

Stratigraphical Succession

The Carboniferous Limestone Series forms wide outcrops on the flanks of the Armorican folds in south Pembrokeshire, Gower, and the Vale of Glamorgan, and it rims the coalfield in a narrow belt along the north and east crops. The lithological sequence is broadly divisible into the three groups of the Lower Limestone Shales, the Main Limestone, and the Upper Limestone Shales, the divisions marking the establishment, the main development, and the conclusion of the major marine cycle represented in the Lower Carboniferous rocks. The lithological divisions, however, do not hold over the whole of South Wales, there being much variation in detailed sequence and in rock facies, and Dixon and Vaughan showed that the

---

[1]For many years, reflecting their similarity to the sequence in the Avon gorge at Bristol, the rocks have been called Avonian. More recently, in wider analogy with the sequence around Dinant in Belgium, they have also come systematically to be called Dinantian, a term having priority over Avonian. The divisions and zones of the Avonian sequence continue, however, to be more readily applicable to the rocks of South Wales than those of the Belgian Dinantian.

strata are more precisely subdivided, on the basis of the fossils found in different types of limestone, into the following zones (*see* Fig. 18):

2. Upper Avonian (wholly Viséan in Dinantian terms)
   Dibunophyllum Zone (D)
   Seminula Zone (S₂)
   Upper *Caninia* Zone (C₂S₁)
1. Lower Avonian (Tournaisian in Dinantian terms except the upper part of the Lower *Caninia* Zone, which is Viséan)
   Lower *Caninia* Zone (C₁)
   Zaphrentis Zone (Z)
   Cleistopora Zone (K)

The Upper Avonian zonal divisions are usually recognizable with comparative ease, although in places the base of the *Seminula* Zone is uncertainly identified. The Lower Avonian divisions are less precise, and over much of the ground the *Zaphrentis* and Lower *Caninia* zones form a unitary group not readily divided.

Cleistopora Zone (K)

The *Cleistopora* Zone is approximately conterminous with the Lower Limestone Shales, rocks consisting of alternations of calcareous shales and thin limestones. Many of the limestones are impure and may weather to rottenstone. Occasional beds are oolitic; some are sandy and quartzitic, or even conglomeratic; a few are stained a hematitic red. The whole zone bears evidence of accumulation in shallow water into which much terrigenous detritus was being brought.

The only corals found in the zone are specimens of *Vaughania* [*Cleistopora*] *vetus* and occasional zaphrentoids. Brachiopods, on the other hand, are common and include *Spirifer tornacensis, Syringothryis cuspidata, Spiriferellina octoplicata, Cleiothyridina* [*Athyris*] cf. *roissyi, Eumetria carbonaria, Camarotoechia mitcheldeanensis, Avonia* [*Productus*] *bassa*, and species of *Chonetes*. The conodonts *Polygnathus, Pseudopolygnathus, Clydagnathus, Siphonodella*, are common in the zone and continue upwards into the *Zaphrentis* Zone. The lowest beds, though not greatly differing in lithology from those above, have a more restricted fauna characterized by the bivalves *Modiolus, Myophoria, Ctenodonta, Grammatodon*, and *Sanguinolites*; and other common fossils are annelids and ostracods. Signs of disturbance by burrowers and scavengers are abundant in some of the beds.

Although the Lower Limestone Shales show little lateral change in lithology they vary greatly in thickness. In the Bosherston district they are at least 600 ft thick, but decrease to 450 ft in the Pembroke syncline at Skrinkle, where they show complete transition from the underlying Devonian Skrinkle Sandstones, and to about 300 ft near Carew. In Gower they are about 500 ft thick, but eastwards they decrease to 350 ft near Bridgend, to 250 ft in the Taff valley, to 120 ft in the Ebbw valley, and to 100 ft on the north-east crop near Abergavenny. (*See* Fig. 19.)

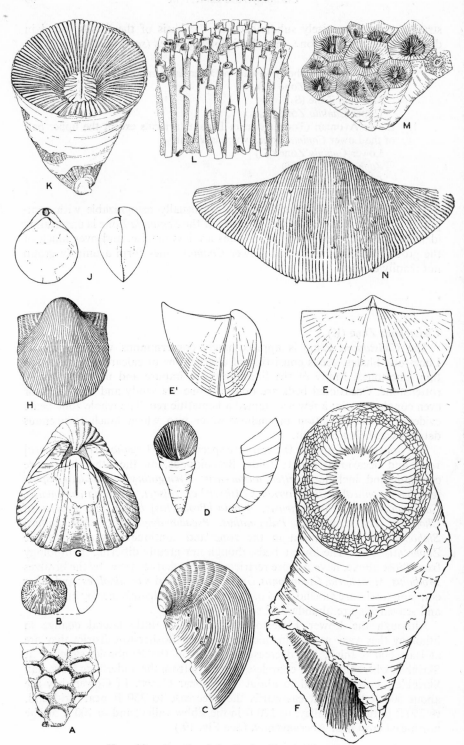

FIG. 18. *Fossils of the Carboniferous Limestone*

Zaphrentis Zone (Z)

Typically, as in south-western Gower and in south Pembrokeshire, the highly fossiliferous crinoidal limestones of the *Zaphrentis* Zone mark the beginning of the 'zaphrentid-phase' deposits that continue into the overlying Lower *Caninia* Zone, and are not easily distinguished from the higher beds. In eastern Gower the zone becomes highly dolomitic and merges into the *laminosa* Dolomites above. Similarly, in the Vale of Glamorgan the beds become dolomitized as they are traced eastwards, and between the Taff and Ebbw valleys they consist wholly of crystalline dolomites and dolomitic mudstones—rock types that also characterize the zone on the north crop near Blorenge. The dolomitization is accompanied by progressive decrease in thickness, from over 500 ft in south-western Gower to 300 ft in eastern Gower and the Bridgend district, to about 150 ft in the Ebbw valley. (*See* Fig. 20.)

Oolites interrupt the sequence of crinoidal limestones at Pendine, in Gower, and in the Vale of Glamorgan (where the Candleston Oolite is a notable member). From Blorenge westwards along the north crop they become dominant in the zone, the beds being called the Oolite Group, and only occasional layers are richly fossiliferous. The oolites, being a sign of shallow waters on a shelf of high evaporation, suggest proximity to a nearby coast, which probably lay at no great distance north of the present north crop.

The fossils of the *Zaphrentis* Zone are closely similar to those of the Lower *Caninia* Zone, except that *Caninia* and *Caninophyllum* are absent, the important corals being the zonal forms *Fasciculophyllum* [*Zaphrentis*] *omaliusi*, *Hapsiphyllum* [*Zaphrentis*] *konincki*, and *Zaphrentites* [*Zaphrentis*] *vaughani*, and species of *Michelinia*. Brachiopods, including *Chonetes vaughani*, *Productus* [*Dictyoclostus*] *vaughani*, *Schuchertella wexfordensis*, *Syringothyris principalis*, and *Spirifer tornacensis*, are abundant especially in the shale bands.

Lower *Caninia* Zone (C₁)

In south-western Gower and the Bridgend district the rocks of the zone fall into two lithological formations:

EXPLANATION OF FIG. 18.

Fossils of the Carboniferous Limestone
(All natural size.)

A. *Vaughania* [*Cleistopora*] *vetus* Smyth; **B.** *Avonia* [*Productus*] *bassa* (Vaughan); **C.** *Dictyoclostus* [*Productus*] *vaughani* (Muir-Wood); **D.** *Hapsiphyllum* [*Zaphrentis*] *konincki* (Milne Edwards and Haime), two views; **E.** *Syringothyris cuspidata* (J. Sowerby), mut. *cyrtorhyncha* North, two views; **F.** *Caninia cylindrica* Scouler; **G.** *Davidsonina* [*Cyrtina*] *carbonaria* (McCoy); **H.** *Linoproductus* [*Productus*] *corrugatohemisphericus* (Vaughan); **J.** *Composita* [*Seminula*] *ficoidea* (Vaughan), two views; **K.** *Dibunophyllum bipartitum bipartitum* (McCoy); **L.** *Lithostrotion junceum* (Fleming); **M.** *Lonsdaleia floriformis* (Martin), forma *crassiconus* McCoy; **N.** *Gigantoproductus* [*Productus*] *latissimus* (J. Sowerby).

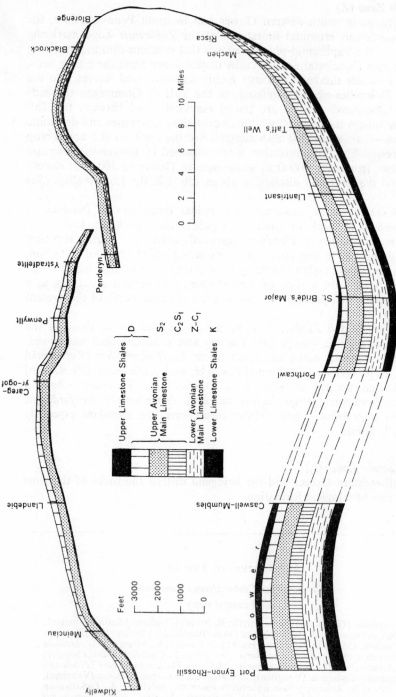

Fig. 19. *Ribbon section of the Carboniferous Limestone flanking the South Wales coalfield*

The reduction in thickness on the north crop is mainly the result of overstep by the *Seminula* Zone across the Upper *Caninia* Zone and the Lower Avonian Main Limestone, on the east crop of overstep by the Millstone Grit almost to the base of the Avonian. The sequence, although remaining thin, is more complete in the north-eastern outcrops by the return of a development of these two zonal groups; but between Blorenge and Risca, along the west flank of the Usk anticline, only a very thin development of lowest Avonian beds remains beneath the Millstone Grit.

2. *Caninia* Oolite: a massively bedded rock, sharply distinguished from the calcite mudstones and crinoidal limestones of the zone above, consisting of a clean white oolite, very well jointed, showing evidence of deposition in shallow water in the occurrence of cross-bedding and contemporary erosion.

1. Crinoidal 'standard' limestones: thinly bedded fossiliferous limestones (sometimes of 'petit granit' type) alternating with calcareous mudstones and shales; the upper beds dark grey or buff finely crystalline secondary dolomites (the *laminosa* Dolomites) with interbedded crinoidal limestones, formed by the dolomitization of normal bioclastic limestones.

Fossils, abundant in the lower formation, include the corals *Michelinia*, *Syringopora*, zaphrentoids, *Caninophyllum* [*Caninia*] *patulum*, and *Cyathoclisia tabernaculum;* and the brachiopods *Dictyoclostus* [*Productus*] *vaughani*, *Leptaena analoga*, *Schizophoria* cf. *resupinata*, *Spirifer tornacensis*, and *Tylothyris* cf. *laminosa*. The *laminosa* Dolomites still contain abundant crinoid debris, but most of the other fossils were destroyed in the process of dolomitization. The *Caninia* Oolite is rich in foraminifers but has few macrofossils, except in thin crinoidal intercalations which contain productids, athyrids, spirifers, and corals of which early clisiophyllid forms (*Koninckophyllum praecursor*), *Palaeosmilia murchisoni*, and *Michelinia grandis* are important. At Three Cliffs Bay in Gower its uppermost beds have yielded the goniatites *Muensteroceras* cf. *inconstans*, *Pericyclus kochi*, and *Prolecanites* cf. *clymeniaeformis*, suggesting equivalence with early Viséan rocks in Belgium.

The separate lithological divisions are not recognizable in the Bosherston outcrops in south Pembrokeshire (*see* Fig. 21) where the greater part of the zone consists of a succession of alternating shales and undolomitized limestones with a rich coral and brachiopod fauna. Moreover, the rocks are with difficulty separated from the overlying crinoid limestones and shales of the Upper *Caninia* Zone, there being no development of the *Caninia* Oolite; and an unbroken sequence of neritic bioclastic sediments accumulated in an off-shore (but not deep-water) environment to a thickness of about 2000 ft.

In the eastern outcrops along the south crop between the Bridgend district and the Ebbw valley, Dixey and Sibly showed that the facies of the *laminosa* Dolomites encroaches upon the overlying and underlying strata (*see* Fig. 20), so that near Cardiff and Newport the whole of the C_1 Zone is an unbroken group of secondarily dolomitized limestones. At the same time the zone is much reduced in thickness from about 500 ft in Gower and near Bridgend to about 350 ft in the Ebbw valley (where it is not readily separated from the underlying dolomites of the *Zaphrentis* Zone). In Pembrokeshire also there is a comparable reduction in thickness as the zone is traced northwards, from about 500 ft near Bosherston to about 400 ft in the Pembroke syncline, and to little more than 200 ft near Tenby. This thinning is likewise accompanied by a development of dolomites and by the incoming of a wedge of *Caninia* Oolite that thickens northwards.

Upper *Caninia* Zone (C₂S₁)

Typically, the Upper *Caninia* Zone consists of highly fossiliferous 'standard' limestones, many crinoidal but some oolitic or porcellanous. The upper beds are locally more oolitic than the lower and may merge upwards without marked break into the thick oolites of the *Seminula* Oolite. Many of the lower strata are thinly bedded and consist of alternations of coarse crinoidal limestones and dark grey calcareous shales crowded with fossils. In the southernmost outcrops of Pembrokeshire (about Bosherston and Linney Head) such fossiliferous limestones (a zaphrentid phase) persist to the top of the zone and even into the basal beds of the overlying *Seminula* Zone. This is the one area in South Wales where reef limestones are known to occur: as Dixon described them, they are massive rocks of poorly bedded calcite mudstone, now dolomitized, in which the only identifiable fossils are fronds of polyzoans and pockets of crinoid debris. They formed mounds on the contemporary sea-floor, the associated sediments, thinly bedded limestones and shales, wrapping round them or interdigitating with their flanks. In many characters the reefs, which occur at two horizons, are similar to the Waulsortian reefs of Belgium and Ireland, although developed on a much smaller scale. In the absence of the *Caninia* Oolite they may be regarded as conveniently defining the local base of the Upper *Caninia* Zone.

In the Pembroke syncline, in south-western Gower, and in the Bridgend district of the Vale of Glamorgan there is an abrupt but otherwise normal junction with the *Caninia* Oolite beneath, but elsewhere, as at West Williamston, in eastern Gower, and in the Miskin district of the Vale of Glamorgan, the upper surface of the *Caninia* Oolite is uneven and slightly eroded, and is followed by a thin group of shales and fine-grained compact porcellanous limestones, the Calcite-Mudstone Group, that locally is the first-formed member of the Upper *Caninia* Zone. The peculiar features of the rocks suggest extremely shallow-water 'lagoonal' conditions of formation; and although the rocks are abundantly fossiliferous, the fossils constitute a peculiar facies in which 'worm'-burrows, faecal pellets, ostracods, and a variety of algal growths (*Mitcheldeania, Ortonella, Garwoodia,* stromatolites) occur in great numbers, but crinoids, brachiopods, and corals are almost completely absent except in occasional thin bands. The group reaches a thickness of 25 ft at Three Cliffs Bay and Caswell in Gower, 40 ft at Tenby, and over 60 ft along the north crop west of Blorenge (where it rests unconformably, with basal grit bands, on the Oolite Group, where it is particularly rich in algal colonies and sheets, and where pseudomorphs after gypsum in a few beds are signs of an evaporitic environment of deposition).

Fossils are exceedingly abundant in the crinoidal limestones of the zone. They include the corals *Caninia cornucopiae, C. cylindrica,* and zaphrentoids in great numbers; a species of *Palaeosmilia* is also common; the compound coral *Lithostrotion* appears for the first time in the Avonian sequence, represented by the species *martini* and *aranea* [*basaltiforme*]; the tabulate corals are represented by *Michelinia grandis* and species of *Syringopora.* Many brachiopods occur, the chief being productids, *Delepinea* [*Daviesiella*] *destinezi, Pustula pyxidiformis, Rhipidomella michelini,* spirifers, *Athyris expansa,* and species of *Composita.* Gastropod beds recur at intervals

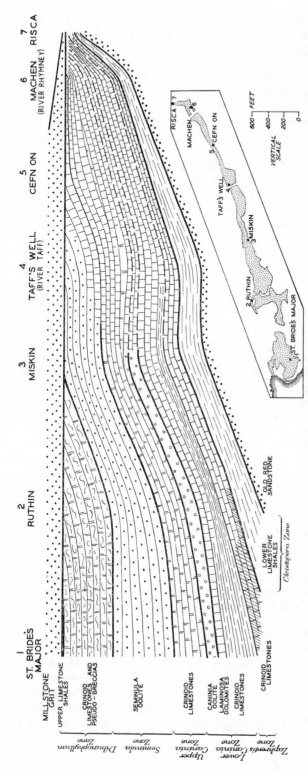

FIG. 20. *Lateral changes in lithology and thickness of the Carboniferous Limestone along the south-east crop of the South Wales coalfield*
The changes are due partly to north-eastward overstep by the Millstone Grit, partly to the transformation of normal limestone into dolomite.
(In part after Dixey and Sibly.)

throughout the zone, some of them crowded with species of *Bellerophon, Euomphalus, Naticopsis, Zygopleura, Loxonema,* and *Aclisina;* associated with the gastropods are specimens of the bivalve *Conocardium.* Many of the beds contain abundant foraminifers. Beds in the lower part of the zone in Gower have yielded well-preserved goniatites (together with nautiloids) referable to *Muensteroceras euryomphalum, M. corpulentum,* and *Merocanites [Prolecanites]* cf. *compressus:* they suggest correlation with lower Viséan rocks in Belgium and Germany. *Cavusgnathus* and *Mestognathus* are significant conodonts which continue upwards into the *Seminula* and *Dibunophyllum* zones.

At maximum thickness the zone exceeds 1000 ft in south Pembrokeshire, but in the Pembroke syncline it is reduced to 400 ft. In Gower also, and in the west of the Vale of Glamorgan, it is no more than 600 ft, though the lithology is much the same as in south Pembrokeshire. Traced eastwards from the Bridgend district along the south crop to the Taff and Ebbw valleys, the zone is still further reduced to some 200 ft near Risca. At the same time there is a great change in lithology, the crinoidal beds of the west being progressively dolomitized, the fossils obliterated, as in the rocks of the zones beneath.

Seminula Zone (S₂)

The most widespread of all the Avonian zones, the *Seminula* Zone, is dominated by massive coarse oolites, the *Seminula* Oolite. Pisolites, algal limestones, and calcite mudstones are intercalated in the upper part and are particularly strongly developed along the north crop. The general characters of the zone suggest deposition in the very shallow waters of a near-surface shelf or bank environment.

Fossils, though common, are concentrated in bands. Chief are the brachiopods *Composita [Seminula] ficoidea* and *Linoproductus [Productus] corrugatohemisphericus (see* Fig. 18), particularly abundant in the upper part of the zone. Corals are represented by species of *Lithostrotion (L. martini* and *L. aranea)* and *Carcinophyllum vaughani.* A characteristic zonal index on the north crop is the brachiopod *Davidsonina [Cyrtina] carbonaria,* which ranges to the top of the zone, but it is relatively rare in the southern outcrops especially in Gower.

The zone is about 800–1000 ft thick in Gower and the Vale of Glamorgan. Along the north crop, however, its thickness does not exceed 400 ft, perhaps because of northward overlap. The thinning is not accompanied by any marked lithological or faunal changes, except that sandy beds appear both at the base (where they may be conglomeratic with large quartz pebbles) and in the upper beds beneath the *Dibunophyllum* Zone.

Dibunophyllum Zone (D)

The beds of the *Dibunophyllum* Zone characteristically consist of neritic crinoid, coral, and brachiopod limestones, many richly foraminiferal, with some oolites. Contemporary shallowing is indicated by the occasional occurrence of thin carbonaceous bands as impure coals. Many of the beds are pseudobreccias—rocks described by Dixon as being the product of the patchy recrystallization (perhaps accompanying some desiccation) of limestones in which a contrast between 'fragments' and 'matrix' is well brought out by differential weathering, and is enhanced locally by a selective dolomitization of the 'fragments'. The topmost beds of the zone (often referred to the Upper *Dibunophyllum* or D_3 Subzone) are represented by the argillaceous phase of the Upper Limestone Shales (the 'Black Lias' of Gower): in places, as at Bishopston in Gower and along the north crop, some of the muddy limestones are decalcified and form rottenstones that have been worked for polishing powder.

Fossils in the zone include an abundance of clisiophyllid corals (*Dibunophyllum, Lonsdaleia, Koninckophyllum*) together with species of *Lithostrotion* and *Palaeosmilia*. Many kinds of brachiopods are also represented, the most common being productoids. The conditions of deposition of the Upper Limestone Shales appear to have been inimical to most of the corals of the underlying 'standard' limestones, and the phase is characterized by the common occurrence of zaphrentoids (*Amplexizaphrentis enniskilleni, A. oystermouthensis, Caninia sp.*) with productoid and spiriferoid brachiopods including *Eomarginifera* [*Productus*] *longispina, Productus concinnus*, pustulids, *Spirifer oystermouthensis, Crurithyris amoena, Martinia multicostata*; and from it *Neoglyphioceras* and *Sudeticeras* are recorded (indicating a P_2 age). In the southern outcrops the upper part of the limestone phase (the Middle *Dibunophyllum* or D_2 Subzone) is not readily separated from the lower beds (the Lower *Dibunophyllum* or D_1 Subzone), but on the north crop it contains *Lonsdaleia floriformis, L. duplicata, Palaeosmilia regia, Orionastrea ensifer*, and species of *Lithostrotion* with small corallites (*Lithostrotion junceum, L. maccoyanum*, and *L. portlocki*), which are absent from the beds beneath. *Palaeosmilia murchisoni, Linoproductus* [*Productus*] *hemisphaericus*, species of *Gigantoproductus*, and locally *Davidsonina* [*Cyrtina*] *septosa* are common forms of the Lower *Dibunophyllum* (D_1) Subzone.

The whole zone reaches a thickness of about 800 ft in Gower, at least 800 ft in the southernmost outcrops in Pembrokeshire, and more than 600 ft in the Bridgend district. In Pembrokeshire and along parts of the main north crop (especially east of the Vale of Neath) the zone is progressively overstepped by the Millstone Grit. Between Llandebie and the Tawe valley, however, and again at Kidwelly and Pendine, all three subzones can be recognized: they show great attenuation, reaching a maximum thickness of less than 300 ft, and a thickness at Kidwelly and Penwyllt of less than 200 ft.

The thinning of the zone along the north crop is accompanied by notable lithological changes, the Lower *Dibunophyllum* Subzone consisting mainly of a massive false-bedded oolite, the Light Oolite, resting on a calcareous sandstone, the Honeycombed Sandstone, that becomes decalcified on weathering and then possesses a cellular structure. Crinoidal limestones and pseudobreccias are well represented only in the western outcrops. Corals are rare in the Light Oolite, the most abundant fossil being *Linoproductus* [*Productus*] *hemisphaericus*, which occurs in profusion. Between the Tawe and Neath valleys seams of sand appear in the Light Oolite, and the texture of the rock may approach that of a grit. The Middle *Dibunophyllum* Subzone is even more arenaceous, few of the limestone beds near Penwyllt and Penderyn being free of sand; it also contains much nodular and tabular chert. Marked faunal changes accompany the development of the sandy facies.

In the southern outcrops of south Pembrokeshire, Gower, and the Vale of Glamorgan, the Middle *Dibunophyllum* Subzone, though containing a rich fauna of corals, is not readily distinguished from the Lower either on lithological or on faunal grounds, for such diagnostic forms as *Lonsdaleia, Orionastraea, Palaeosmilia regia*, and *Productus productus* are rare or absent. At Kidwelly the subzone presents the same crinoidal and pseudobrecciated facies, free of sand, as it does in Gower a few miles to the south; at the same time it possesses a richer fauna than the equivalent beds in Gower, and the usual species of *Dibunophyllum* and productoids are accompanied by the *Lonsdaleia* fauna. At Crwbin, some four miles north-east of Kidwelly, the subzone is very different in thickness, lithology and fossils. The 80 ft of 'standard' limestones near Kidwelly are replaced by over 200 ft of limestones with sandy and conglomeratic layers, coarse brecciated beds, and calcite mudstones; and the abundant corals and brachiopods of the western outcrop are, with the exception of the species of *Productus*, almost wholly absent, their place being taken by molluscs (gastropods and bivalves), notably *Sanguinolites contortus* and species of *Aviculopecten*. Farther east, towards Penwyllt, the lithology remains siliceous, but the contained fauna becomes richer and in the easternmost outcrops is similar to that near Kidwelly, containing clisiophyllid corals together with many zaphrentoids (*Amplexizaphrentis enniskilleni, Caninia juddi*).

Conditions of Deposition

The details of the Avonian sequence may be summarized in the statement that the major marine cycle of sedimentation preserved in the rocks is composed of three minor cycles. The first begins with the shallow-water muddy and sandy strata of the Lower Limestone Shales; its clearer-water phase is expressed in the crinoid and coral limestones of the *Zaphrentis* and the Lower *Caninia* zones; its close is marked by the false-bedded *Caninia* Oolite. The second, beginning with the Calcite-Mudstone Group, is chiefly displayed in the fossiliferous zaphrentid-phase deposits of the Upper *Caninia* Zone, which are followed by the oolites and algal limestones of the *Seminula* Zone. The coral and shelly limestones and pseudobreccias of the *Dibunophyllum* Zone, closing with the Upper Limestone Shales, mark the third.

Lateral variation in thickness is consistently one of attenuation towards the north in Pembrokeshire and towards the north and north-east in Glamorgan and Monmouthshire. The thinning is almost invariably accompanied by lithological changes. In the Upper Avonian rocks the pure limestones of the south are replaced by sandy and even conglomeratic beds on the north, of which the Honeycombed Sandstone of the Lower *Dibunophyllum* Subzone is an example. In the Lower Avonian rocks and in the Upper *Caninia* Zone, the principal development in the north-eastern outcrops is of dolomites and oolites. It is evident that the deeper waters of maximum sedimentation lay towards the south, and the contemporary fluctuating shore-line, sometimes with wide lagoonal flats, sometimes with streams bringing in much terrigenous sand and mud, lay not far to the north of the present north crop. (*See* Fig. 21.)

Mid-Avonian Unconformities

In the southern outcrops the Upper *Caninia* Zone rests on the *Caninia* Oolite with a marked change in lithology, but without other indication of a break in the succession. But in eastern Gower, and, to a still greater degree, at West Williamston on the Carew anticline, there is evidence that the Oolite was eroded, pitted, and piped before the overlying Calcite-Mudstone Group was laid down. It is clear, therefore, that the uplift that closed the minor cycle of deposition with the *Caninia* Oolite was at these places sufficient to cause emergence of the sea-floor above sea-level. Still farther north, nearer the massif of St. George's Land, uplift caused the rocks already formed to suffer extensive erosion and removal. The Upper Avonian stage then rests with strong unconformity on the residual Lower, successive beds overlapping and overstepping progressively northwards. (*See* Fig. 21.)

The complete geological history cannot be precisely determined, because much of the evidence is now hidden beneath the Upper Carboniferous rocks of the coalfield, but sufficient remains both in Breconshire and in Pembrokeshire to reveal the main events. Along the north-east crop, from Blorenge to the Vale of Neath, the Calcite-Mudstone Group of the Upper *Caninia* Zone rests directly on a residual Oolite Group belonging to the *Zaphrentis* Zone, the Lower *Caninia* Zone being overstepped. The unconformity is marked by the occurrence in the Calcite-Mudstone Group of conglomerates containing pebbles, some of them algal-filmed, derived from the Oolite Group, and by the transgressive descent westwards in present outcrops of the Calcite-Mudstone Group across successively older beds of the Oolite Group until at the limit of outcrop only some 60 ft of Lower Avonian limestones remain between the Upper Avonian rocks and the Lower Limestone Shales. In Pembrokeshire, Sullivan has amplified Dixon's work in showing that similar changes take place northwards, the unconformity seen in its first stages on the flanks of the Sageston anticline, where the Calcite-Mudstone Group rests on a pocked and eroded surface of the *Caninia* Oolite, being intensified at Pendine, where only 70 ft of the *Zaphrentis* Zone remain beneath the Calcite-Mudstone Group, which, strongly conglomeratic, descends in deep channels into its eroded surface.

The unconformity beneath the Upper *Caninia* Zone is a major factor in explaining the great reduction in thickness of the Lower Avonian sequence as it is followed northwards in South Wales; and at original depositional limits, which lay not very far north of the present north crop, the Lower Avonian rocks may well have been completely extinguished by the transgressive Upper *Caninia* Zone. Deposition of the Upper *Caninia* Zone was followed by a second pulse of uplift and erosion, as shown by the unconformity, scarcely of less magnitude than the earlier, at the base of the *Seminula* Zone. The signs are clearly shown along the north-east crop (*see* Fig. 19), where the Upper *Caninia* Zone, at maximum only 60 ft thick and preserved only in its lowest member, is overstepped westwards until at a feather-edge near the Tawe valley it finally disappears, the beds of the

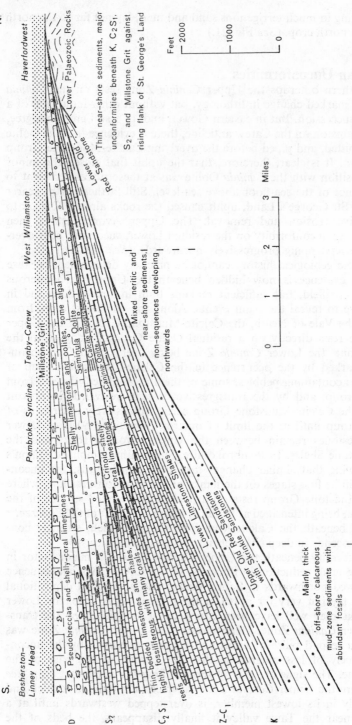

FIG. 21. *Reconstructed section of lateral changes in the Carboniferous Limestone of Pembrokeshire*

The reduction in thickness northwards is a product of multiple overstep—of the Lower Old Red Sandstone by the Upper, of the Lower Avonian rocks by C_2S_1, of C_2S_1 by S_2, and of the whole of the Avonian by the Millstone Grit. The thinning is a reflection of the northern rise of St. George's Land in relation to subsidence in southernmost Pembrokeshire. There are corresponding facies changes in the sediments, notably in the replacement by fossiliferous neritic limestones in the south by oolites and algal limestones in the north.

(In part after Sullivan.)

A. Glacial spillway at Dinas Head near Fishguard

(Cambridge Univ.)

B. The coast of south-west Gower

(Cambridge Univ.)

South Wales *(Geol. Surv.)*

Plate VI

Seminula Zone descending first on to the residual Oolite Group and then on to the Lower Limestone Shales. A similar development is to be seen along the north crop in Pembrokeshire: the Upper *Caninia* Zone, the underlying residual *Zaphrentis* Zone seen at Pendine, the Lower Limestone Shales, and the Old Red Sandstone are progressively overstepped until the *Seminula* Zone comes to rest directly on Silurian rocks near Haverfordwest. *See* Fig. 21.)

In summary, it may thus be said that there is a thick and complete sequence of Avonian rocks in southern outcrops, but that northwards, as the flanks of St. George's Land are approached, the sequence is reduced in consequence of mid-Avonian movements, partly by the thinning of individual members, but mainly by the intensification of two unconformities, as a combined result of which the Upper *Caninia* Zone and all the Lower Avonian zones are extinguished.

References

DIXEY, F. and SIBLY, T. F. 1918. The Carboniferous Limestone series on the south-eastern margin of the South Wales coalfield. *Quart. J. Geol. Soc.*, **73**, 111–64.

DIXON, E. E. L. and VAUGHAN, A. 1912. The Carboniferous succession in Gower (Glamorganshire), with notes on its fauna and conditions of deposition. *Quart. J. Geol. Soc.*, **67**, 477–571.

EVANS, W. D. and COX, A. H. 1956. An Old Red Sandstone–Carboniferous Limestone series junction at Tongwynlais, north of Cardiff. *Geol. Mag.*, **93**, 431–4.

GEORGE, T. N. 1927. The Carboniferous Limestone (Avonian) succession of a portion of the north crop of the South Wales coalfield. *Quart. J. Geol. Soc.*, **83**, 38–95.

—— 1928. The Carboniferous outlier at Pen-Cerig-calch. *Geol. Mag.*, **65**, 162–9.

—— 1933. The Carboniferous Limestone series in the west of the Vale of Glamorgan. *Quart. J. Geol. Soc.*, **89**, 221–72.

—— 1939. The Cefn Bryn shales of Gower. *Geol. Mag.*, **76**, 1–6.

—— 1952. Tournaisian facies in Britain. *Rep. 18th Int. Geol. Congr.* (Gt. Britain), **10**, 34–41.

—— 1954. Pre-Seminulan Main Limestone of the Avonian series in Breconshire. *Quart. J. Geol. Soc.*, **110**, 283–322.

—— 1956. Carboniferous Main Limestone of the east crop in South Wales. *Quart. J. Geol. Soc.*, **111**, 309–22.

—— 1956. The Namurian Usk anticline. *Proc. Geol. Assoc.*, **66**, 297–316.

—— 1958. Lower Carboniferous palaeogeography of the British Isles. *Proc. Yorks. Geol. Soc.*, **31**, 227–318.

—— 1969. British Dinantian stratigraphy. *C. R. 6me Congr. Strat. Carb.*, **1**, 193–218.

—— and HOWELL, E. J. 1939. Goniatites from the *Caninia* Oolite of Gower. *Ann. Mag. Nat. Hist.*, (11), **4**, 545–61.

—— and PONSFORD, D. R. A. 1935. Mid-Avonian goniatites from Gower. *Ann. Mag. Nat. Hist.*, (10), **14**, 354–70.

KELLING, G. and WILLIAMS, B. P. J. 1966. Deformation structures of sedimentary origin in the Lower Limestone Shales (basal Carboniferous) of south Pembrokeshire. *J. Sed. Pet.*, **36**, 927–39.

OWEN, T. R. and JONES, D. G. 1955. On the presence of the Upper *Dibunophyllum* zone (D_3) near Glynneath, South Wales. *Geol. Mag.*, **92**, 457–64.

ROBERTSON, T. and GEORGE, T. N. 1929. The Carboniferous Limestone of the north crop of the South Wales coalfield. *Proc. Geol. Assoc.*, **49**, 18–40.

SIBLY, T. F. 1920. The Carboniferous Limestone of the Cardiff district. *Proc. Geol. Assoc.*, **31**, 76–92.

SULLIVAN, R. 1965. The mid-Dinantian stratigraphy of a portion of central Pembrokeshire. *Proc. Geol. Assoc.*, **76**, 283–300.

—— 1966. The stratigraphical effects of the mid-Dinantian movements in south-west Wales. *Palaeogeogr., Palaeoclimatol., Palaeoecol.*, **2**, 213–44.

THOMAS, T. M. 1953. New evidence of intraformational piping at two separate horizons in the Carboniferous Limestone (D₂) at South Cornelly, Glamorgan. *Geol. Mag.*, **93**, 73–82.

8. Millstone Grit

Sub-Namurian Unconformity

Yet a third unconformity had major effect on the Avonian sequence. Its origin, as of the mid-Avonian unconformities, is to be ascribed to uplift along the southern flanks of St. George's Land, to southward migration of the contemporary shores, and to erosion of the exposed Avonian rocks. The unconformity lies beneath the Millstone Grit, and reflects movements of earliest Namurian age. At its extreme expression, near Haverfordwest, the Avonian rocks are completely overstepped, the Millstone Grit descending on to Lower Palaeozoic, perhaps Pre-Cambrian, rocks. Details of the unconformity are sufficiently well preserved in present outcrops to allow a reconstruction of the tectonic form of the surface on which the Millstone Grit was deposited: they are illustrated in Fig. 22, in which boundary lines of the sub-Namurian outcrops of the Avonian zones serve as contour-lines to define structure.

It is clear that in the central and western parts of the province overstep was towards the north, in continued reflection of Avonian tectonics. In the east, however, a variant structure is emphasized by the north-and-south alignment of the zonal junctions. It reaches maximum intensity along the north-east crop, where in places the Millstone Grit descends on to the Lower Limestone Shales, which, only 100 ft thick, are the last remnants of the Avonian sequence. It is significant that immediately to the east lies the Usk anticline with Silurian rocks flanked by Old Red Sandstone in its core: although the fold, lying between the synclinal coalfields of South Wales and the Forest of Dean, is mainly post-Carboniferous (Hercynian) in development, it was thus already strongly expressed in early Namurian times (and was to show repeated pulsatory growth during later Namurian and Westphalian times also—*see* p. 94).

Near Kidwelly, the Penlan Quartzite, a rock of Millstone-Grit lithology proved by its fossils to be of Carboniferous age, lies on a floor of Old Red Sandstone. If it is in its original position, as Dixon and Pringle supposed, it indicates the axis of a similar sharp pre-Namurian fold; and it has been taken to demonstrate the local intensity of Namurian overstep, for immediately to the east the Avonian sequence includes the *Dibunophyllum* Zone up to the Upper Limestone Shales. Alternatively the Quartzite may rest only accidentally on the Old Red Sandstone, having been let down by solution subsidence in times long post-Carboniferous; and it then compares in position with similar masses of Millstone Grit out of stratigraphical context that T. M. Thomas has described as occurring at many places along the north crop.

Lithology and Succession

The sub-Namurian unconformity is a sign of the radical changes in the environment of sedimentation brought about by the earth-movements in mid-

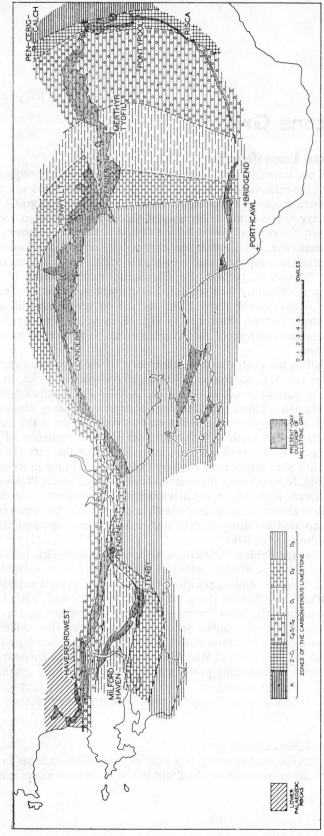

Fig. 22. *Map of the foundation on which the Millstone Grit was deposited*

The disappearance northwards of the Avonian zones is a reflection of the approach to the west flank of the Usk anticline. The broken outcrops between Penwyllt and Merthyr Tydfil are a consequence of movement along the Neath disturbance, which did not come into being on such a scale until after the deposition of the Millstone Grit.

PRESENT-DAY OUTCROP OF MILLSTONE GRIT

ZONES OF THE CARBONIFEROUS LIMESTONE

| K | Z–C_1 | C_2S_1–S_2 | D_1 | D_2 | D_3 |

LOWER PALAEOZOIC ROCKS

0 1 2 3 4 5   10 MILES

Carboniferous times. The strata succeeding the Carboniferous Limestone are sharply distinguished lithologically from the underlying clear-water typically marine, shelly and coral limestones, and consist of a varied and variable group of sandstones and shales. Many of the beds show rapid lateral changes from fine silty muds to coarse sandstones and grits and even to conglomerates. Most of the coarser rocks are wedge-bedded and current-bedded, and few horizons, except in a broad lithological sense, can be traced over the whole of South Wales. The nature of many of the sediments thus strongly suggests conditions of accumulation characteristic of the estuaries of large rivers, in which fast currents loaded with terrigenous detritus deposit their burden in lenticular aprons and festoons; any one bed of such a series has no great lateral uniformity or persistence but as a whole the series of deposits may present the same broad appearance over many square miles.

The coarser rocks are white or yellow quartzites and quartz conglomerates containing grains and pebbles often cemented in a siliceous matrix. In general appearance, apart from their colour, they resemble some of the grits and conglomerates of the upper beds of the Old Red Sandstone, and in large part they appear to have been derived from a land-mass of Pre-Cambrian and Lower Palaeozoic rocks and Old Red Sandstone lying to the north. The quartzitic rocks, of which the Basal Grit is the chief formation, form hard bands strongly resistant to denudation and give rise to escarpments surrounding the coalfield. Beyond the Carmarthenshire Fans, where the Upper Old Red Sandstone thins westwards, the Basal Grit takes its place and forms the protective element capping the multiple scarp face; and even along the east crop, where the Quartz Conglomerates of the Old Red Sandstone are well developed, the Namurian quartzites and grits mark the scarp summit.

In detail the succession within the Millstone Grit often shows rhythms of sedimentation. Beginning with a coal seam (rarely of economic importance and often little more than a black carbonaceous band), a complete cycle shows a sequence of changes through grey or black fine-grained marine shales often richly fossiliferous with molluscs and brachiopods, sandy shales and siltstones with bedding often disturbed by 'worm' burrows (notably *Planolites*) and other scavenger tracks, and fine-grained grits that merge upwards into quartz conglomerates. The conglomerates may show evidence of being covered by an old soil or rootlet bed, and are overlain by another coal streak beginning the succeeding cycle.

Along the north crop a comparable generalized cycle characterizes the Millstone Grit as a whole, and prompted early division of the series into three broad lithological formations:

3. Farewell Rock, so called because, once reached in mines, it indicates little likelihood of workable coal being found beneath: brown, green, and yellow sandstones and quartzites.

2. Shale Group or Middle Shales: dark grey, blue, and black shales with bands of grit, sandstone, and quartzite. The Twelve-Foot Sandstone near the base is a notable member.

1. Basal Grit: a very variable group of quartz-grits, quartzites, sandstones, and conglomerates, some of the beds very pure (with over 99 per cent quartz) and worked for fire-brick (silica brick). Bands of shale, thin and

insignificant in the lower part, become increasingly abundant upwards, so that the division passes by stratal alternation into the overlying Shale Group.

In general, the sediments were derived from the north and were carried by powerful streams southwards. The contemporary land-mass and the 'shore-line' or coastal margins lay, as in Avonian times, not far to the north of the present outcrops, and deeper waters extended over the (present) Bristol Channel region into Devon and Cornwall. As a whole, the Millstone Grit shows increasing fineness of grain when traced southwards away from the shore-line, and in southern Pembrokeshire, in Gower, and in the Bridgend district the series is dominated by shales in which thin impersistent sandstones, quartzites, and grits are relatively unimportant. While therefore a lithological grouping of the rocks is still conveniently used in generalized description, the individual members prove to be highly variable in development, not everywhere identifiable, and diachronous when followed laterally. Along the north crop, where the grouping is best seen, the Basal Grit encroaches upwards in zonal sequence towards the east; the Shale Group is correspondingly reduced; and 'Farewell Rock' was a name misleadingly ascribed to sandstones both in the uppermost Millstone Grit and in the lowest Coal Measures. A more precise means of correlation is provided by the fossils, the lithologically named Millstone Grit then being equated with the palaeontologically defined Namurian.

Fossils

With the great change in conditions of deposition from the Avonian limestones to the Namurian grits and shales, there was a rapid extinction or emigration of the abundant corals and brachiopods characteristic of the Avonian rocks, and these forms, except in thin impure calcareous bands, are absent from most of the Millstone Grit. Their place is taken by organisms adapted to life on a muddy sea-floor, chiefly bivalves and goniatites (*see* Fig. 18). Even these are found only at certain horizons in the shales, the sandstone and grits being almost wholly unfossiliferous, though drifted plant remains (*Lepidodendron, Sigillaria, Stigmaria*) may occur even in the coarsest conglomerates.

The invertebrate fauna, relatively rich and varied in bands of calcareous shales laid down in the shallower nearshore waters towards the margins of the depositional basin, includes bivalves (scallops, myas, nuculids), gastropods, brachiopods (orthoids, chonetids, spirifers, productoids, athyrids, the relatively common mud-burrowing *Lingula*, and *Orbiculoidea* cf. *nitida*), occasional bryozoans, and rare zaphrentoid corals. Some forms, particularly in the upper part of the Grit where they are found in shales associated with thin coal seams, are freshwater and brackish-water mussels (*Carbonicola lenicurvata, Anthraconaia* cf. *bellula*) that link the series with the overlying Coal Measures and mark the approach of more typically swamp conditions. Many shale bands contain abundant relics of the worm *Planolites*.

The goniatites (*see* Fig. 23), concentrated in widespread marker beds, are of particular stratigraphical importance. By their aid it is possible to apply

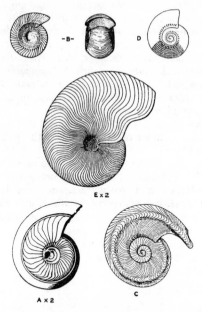

FIG. 23. *Upper Carboniferous goniatites*
(All natural size except where stated.)

A. *Eumorphoceras bisulcatum* Girty, E_2, Millstone Grit; **B.** *Homoceras bey-richianum* (Haug), two views, H, Millstone Grit; **C.** *Reticuloceras super-bilingue* Bisat, R_2, Millstone Grit, (diagrammatic reconstruction); **D.** *Gastrioceras cancellatum* Bisat, G, Millstone Grit, (diagrammatic reconstruction); **E.** *Politoceras politum* (Shumard) [*Homoceratoides jacksoni* Bisat], Cefn Coed Marine Band, in *similis-pulchra* Zone, Coal Measures.

Bisat's zonal scheme, first devised in the Pennines, to South Wales, all the major divisions being recognized:

4. The *Gastrioceras* Stage (G_1) (the overlying G_2 continuing upwards in the Coal Measures from its definitive base, the *Gastrioceras subcrenatum* Band). The stage contains only a few layers with common fossils: its base lies at a notable and widespread marker bed, the *Gastrioceras cancellatum* Band. At a higher horizon the *Gastrioceras cumbriense* Band is also a useful marker. Other goniatites occurring in the stage are *Gastrioceras crenulatum*, *G. crencellatum*, *Agastrioceras carinatum*, *Homoceratoides divaricatus*, and *Anthracoceras spp.* Bivalves include *Aviculopecten aff-losseni*, *Dunbarella elegans*, *Nucula aequalis*, *Polydevcia acuta*, *P. stilla*, *Palaeoneilo* cf. *laevirostris*, *Edmondia lowickensis*, *Parallelodon reticulatus*, and *Posidonia gibsoni*. Spirifers are relatively common (with *Spirifer sp.*, *Martinia sp.*, and *Crurithyris carbonaria*), and are found with *Lissochonetes geinitzianus*, *Schizophoria hudsoni*, *Rhipidomella* cf. *michelini*, and *Productus carbonarius;* and fenestrate bryozoans are recorded.

3. The *Reticuloceras* stages (R_2 and R_1). The upper stage, R_2, although it reaches thicknesses of hundreds of feet, is poorly represented by fossil beds, the *Reticuloceras superbilingue* Band and the *Anthracoceras* Band alone being widely developed within it. Its base can thus be identified only approximately. Other fossils in the stage include *R. bilingue,*

Donetzoceras sigma, a late *Homoceras* (*striolatum*) and an early *Gastrio-ceras* (aff. *lineatum*); and *Sanguinolites ovalis, Nuculopsis gibbosa, Polidevcia* [*Nuculana*] aff. *acuta*, and *Productus carbonarius* also occur.

The lower stage, R_1, is characterized by a number of fossiliferous beds containing *Reticuloceras reticulatum, R. eoreticulatum, R. circumplicatile*, and *Homoceratoides mutabilis*, with *Schizophoria hudsoni, Dunbarella rhythmica, Caneyella squamula, Posidonia obliquata, Posidoniella* cf. *minor* and sponges (*Hyalostelia*). The *R. reticulatum* and *R. circumplicatile* bands are of some importance as markers.

2. The *Homoceras* Stage (H). The development of the stage is usually thin, but it contains a number of fossil bands with *Homoceras beyrichianum*, and *H. subglobosum*.

1. The *Eumorphoceras* stages (E_2 and E_1). The upper stage, E_2, is not well represented by goniatites, but *Cravenoceratoides stellarum, Eumor-phoceras bisulcatum* and *Nuculoceras nuculum* have been recorded, together with *Posidonia* cf. *corrugata* which also may be indicative. A plant association, notably including *Lyginopteris stangeri*, appears to be characteristic.

The lower stage, E_1, has not been recognized at most outcrops, pre-sumably being widely overlapped, but *Eumorphoceras pseudobilingue* has been recorded in Gower.

Lateral Variation

In the northern outcrops the fullest Namurian development is found between the Loughor and Tawe valleys, where the broad sequence of stages, as described by Ware, is complete (E_1 probably excepted) in a thickness of about 800–900 ft. The Upper Limestone Shales are suspiciously thin, but are followed without obvious break by the Plastic Clay beds of E age, uniformly fine quartzitic clays formerly worked for polishing powder. They are overlain by some 300–400 ft of Basal Grit that includes thin shale bands with H and R_1 goniatites. The Shale Group, dominating the upper part of the sequence, spans R_2 and G_1: it is capped by thin 'Farewell Rock', and includes in R_2 a prominent rib of hard grit, the Twelve-Foot Sandstone, that forms a persistent marker bed.

Westwards the thickness is reduced to about 600 ft near Kidwelly, where the E and H stages are greatly condensed, and the Basal Grit with them. Along the north crop in Pembrokeshire the Basal Grit is insignificantly developed, and although the full thickness remains at about 550 ft the zonal representatives, as Archer has demonstrated, are restricted to the R and G stages: the contact at Haroldston St. Issell's near Haverfordwest of R_2 grits with limestones of the *Seminula* Zone, as reported long ago by O. T. Jones, remains a classical indication of the unconformity developed between the Lower and the Upper Carboniferous rocks.

Eastwards the changes in the Namurian sequence are more rapid and more radical. The thickness is reduced to about 450 ft in the Neath valley, to 300 ft in the Taff valley, to 150 ft near Brynmawr and less than 100 ft a mile to the east, and perhaps (as Jones and Owen interpret the sequence) to nil (by Ammanian overstep) in the extreme north-eastern outcrops near Blorenge. The reduction is accompanied by an upward extension of the Basal-Grit lithology into R_2 and G_1, and by the disappearance of thick

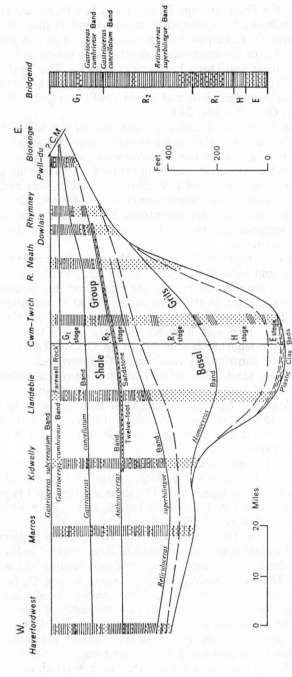

FIG. 24. *Sections of the Millstone Grit between Pembrokeshire and Monmouthshire, to show progressive thinning particularly eastwards against the flank of the Usk anticline*

The lateral changes include overlap of the E and H stages by the R stages, a diachronous migration of the Basal Grit, and a regional development of grits in the north and shales in the south.

(In part after Archer, Ware, Robertson, Jones and Owen, and Woodland.)

sandstones above the Shale Group. D. G. Jones has traced a very thin remnant of E_2 (indicated by *Lyginopteris stangeri*) and H into the Neath valley, but thereafter in Robertson's description R overlaps and oversteps on to Avonian rocks. In the easternmost development the basal sediments may according to Jones and Owen be no older than R_2. The Twelve-Foot Sandstone is identifiable as far as the Taff valley, perhaps as far as the Clydach valley, but beyond it merges with and is not readily separable from the diachronous Basal Grit. (*See* Fig. 24.)

A particular feature of the increasing magnitude of the sub-Namurian unconformity eastwards—Avonian zones progressively overstepped, lower Namurian zones overlapped—is the occurrence of puzzling quartz-pebble conglomerates in what appear to be the uppermost Avonian beds present (of no matter what zonal age down to Seminulan), which Owen and Jones have ascribed to a highly diachronous onset of a 'Millstone-Grit' facies while the limestones were still not completely lithified, and to a thinning of the Namurian sequence less by simple overlap than by slow deposition with repeated non-sequence.

Along the east crop southwards to Risca the series remains very thin, goniatites are rare, and beds of R_2 age lie at the base. Farther south, few details are known as the rocks swing into the south crop, but Squirrell has recorded a thickness of 100 ft in the Ebbw valley, of 140 ft near Rudry, and of 175 ft in the Taff valley. There is great expansion to 900 ft towards Bridgend, where Woodland and Ramsbottom have shown that although the greater part of the sequence is referable to the G_1 and R_2 stages (with the marker bands of *Gastrioceras cumbriense, G. cancellatum,* and *Reticuloceras superbilingue* identifiable, as they are for many miles along the north crop), the R_1, H, and E stages return and are rich in fossiliferous horizons. Shales dominate the sequence, and while there are many lenticular beds and ribs of sandstone, it is not possible to distinguish the Basal Grit or the Twelve-Foot Sandstone as well-defined members, and a Namurian Farewell Rock does not occur.

The sequence in Gower, described by Dix with much further detail supplied by Stephens, is exceptionally thick, nearly 2300 ft, and although the manner of the lateral passage is unknown there must be rapid expansion westwards across Swansea Bay. E (of which both E_1 and E_2 are represented), reaches nearly 600 ft, H is nearly 200 ft, and R_1 and R_2 together are about 1300 ft: only G_1, at about 100 ft, is usually thin. The sequence is dominated by shales but there are thick sandstones, some of them massive, in R_2. In the Tenby outcrops, where folding and faulting confuse a reading of the stratigraphy but where H (with *Homoceras beyrichianum*), R, and G_1 (with the *Gastrioceras cumbriense, G. cancellatum,* and *Anthracoceras* bands) have been identified by D. G. Jones, the sequence also is dominantly of shales with a number of fossil horizons, and probably exceeds 1000 ft in thickness.

The lateral changes clearly reflect the double controls of the rising massif of St. George's Land to the north, and the emerging arch of the Usk anticline to the east, which together formed the flanks of a marked embayment, with steeply falling floor, whose axis ran northwards between the Loughor and Tawe valleys (*see* Fig. 25). In the embayment the Namurian rocks of the north crop (notably the Basal Grit) are unusually thick and include

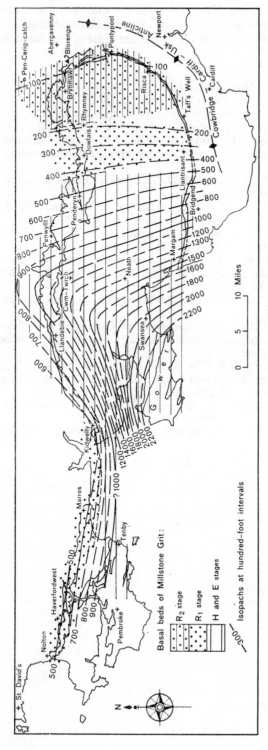

FIG. 25. *Isopach map of the Millstone Grit*

The progressive thinning northwards towards St. George's Land is well shown in the western outcrops, and eastwards towards the Usk anticline in the intervening embayment is particularly well defined by the 900-ft isopach.

representatives of E and H. It is significant that the embayment is also
revealed by the pattern of Namurian overstep across Avonian zones (*see*
Fig. 22), and by the isopachs of the *lenisulcata* Zone of the Coal Measures.

References

ARCHER, A. A. 1965. Notes on the Millstone Grit of the north crop of the Pembrokeshire coalfield. *Proc. Geol. Assoc.*, **76**, 137–50.

DIX, E. 1931. The Millstone Grit of Gower. *Geol. Mag.*, **68**, 529–43.

—— 1933. The succession of fossil plants in the Millstone Grit and the lower
portion of the Coal Measures of the South Wales coalfield (near Swansea)
and a comparison with that of other areas. *Palaeontographica*, **78**, 158–
202.

DIXON, E. E. L. and PRINGLE, J. 1927. The Penlan Quartzite. *Sum. Prog. Geol.
Surv.* for 1926, 123–6.

EVANS, D. G. and JONES, R. O. 1929. Notes on the Millstone Grit of the north
crop of the South Wales coalfield. *Geol. Mag.*, **68**, 164–77.

JONES, D. G. 1969. The Namurian succession between Tenby and Waterwynch,
Pembrokeshire. *Geol. J.*, **6**, 267–72.

—— and OWEN, T. R. 1957. The rock succession and geological structure
of the Pyrddin, Sychryd, and Upper Cynon valleys, South Wales. *Proc.
Geol. Assoc.*, **67**, 232–50.

JONES, O. T. 1925. The base of the 'Millstone Grit' near Haverfordwest. *Geol.
Mag.*, **62**, 558–9.

OWEN, T. R. 1964. The tectonic framework of Carboniferous sedimentation in
South Wales. *In* L. van Straaten (editor): *Developments in sedimentology*,
1, 301–7. Amsterdam.

—— and JONES, D. G. 1961. The nature of the Millstone Grit-Carboniferous
Limestone junction of a part of the north crop of the South Wales coalfield.
Proc. Geol. Assoc., **72**, 239–49.

—— —— 1967. The Millstone Grit succession between Brynmawr and Blorenge.
Proc. Geol. Assoc., **77**, 187–98.

WARE, W. D. 1939. The Millstone Grit of Carmarthenshire. *Proc. Geol. Assoc.*,
50, 168–204.

9. Coal Measures

Lithology

The Coal Measures of South Wales consist almost wholly of terrigenous detritus derived from nearby sources and carried into a shallow subsiding trough of sedimentation by rivers from a land-mass lying, as in Avonian and Namurian times, mainly to the north. The deposits appear to have been laid down under estuarine or freshwater conditions as alluvial muds and sands, swamp clays, and delta fans and aprons; the coarser layers are often cross-bedded in lenticular units; and channelled washouts are common. Marine beds are few and thin, and are restricted to the lower part of the sequence. The region throughout the period of accumulation of the Coal Measures appears never to have been at any great height above sea-level, or to have been isolated as a land-locked basin, and the sediments are paralic. The recurrent coal-seams were formed as thick beds of water-logged peaty humus in swamps and marshes supporting a luxuriant vegetation, through which tributaries and distributaries of large rivers meandered.

Sedimentation of the Coal Measures, as of the Millstone Grit, was in rhythmic cycles. In each typical cycle (cyclothem) a coal-seam, formed at or very near water-level as a thick and dense peat, is succeeded by fine-grained shale or an impure limestone, with goniatites, bivalves, gastropods, brachiopods, foraminifers, even occasional corals, as a sign of sharp subsidence and of a marine incursion or the influence of one. There follow shales increasingly shallow-water and non-marine in character, with a fauna of mussels, which in turn gradually pass upwards into sandy shales, sandstones, and sometimes grits and conglomerates. The close of the cycle is marked by very shallow-water muds, now converted to underclays and rootlet beds by the growth of forests on them, which are overlain by a coal-seam initiating the next cyclothem. Usually the full cyclothem is not represented, one or more elements being omitted; nor is there a constant regularity in the proportionate thickness of the elements.

In many of their features, therefore, the rocks of the Millstone Grit and the Coal Measures are very much alike, and the two series, only arbitrarily distinguished from one another, are aptly combined as Upper Carboniferous. The main differences between them are that the Millstone Grit contains few and mostly unworkable coal-seams, and that the Coal Measures contain only few and thin marine bands; and that in the same kind of sedimentary terrain the Millstone Grit was deposited mainly in a marine delta or prodelta environment, the Coal Measures in a brackish or freshwater delta environment.

Over a great part of the coalfield the Coal Measures as a major lithological suite are dominated by shales in the lower part and by sandstones in the upper. Residual outliers in the deeper synclines preserve above the main mass of sandstones a group of shales in which intercalated sandstones are relatively minor. Moreover, there is a concentration of the richer coal-seams in the lower part of the sequence, to which also the thin marine bands are confined. A ready lithological classification of the rocks that generalizes these stratal relations is one that impressed itself on 19th-century geologists and was embodied in a threefold grouping of the measures into:

3. Upper Coal Series or Supra-Pennant Series, predominantly shales.
2. Pennant Series, predominantly sandstones, of which the massive Pennant Sandstone is the chief formation.
1. Lower Coal Series, predominantly shales.

With increasingly precise identification and correlation of coal-seams, however, and with the tracing of individual beds of shale and sandstone across the coalfield, it has become obvious during recent years that, like their analogues in the Millstone Grit, the lithologically identified members of the Coal Measures change markedly in their characteristics and their thickness as they are followed laterally and are unreliable or deceptive as guides in correlation. In particular, the Pennant Sandstone is found to be especially variable: its base is strongly diachronous, and it is much more massively developed in the west than in the east. A more accurate and sustained correlation is possible when the crude lithological divisions are discarded (or are used only for informal convenience) and are replaced by more refined divisions established by the use of precisely identified marker horizons (notably the coal seams) and of fossils.

Marine Bands

In the confusion resulting from a naïve lithological correlation of the Coal Measures, attempts to use fossils and their vertical distribution as anchors of reference have centred mainly on the non-marine mussels and on the plants—groups that are relatively widespread but that nevertheless are not always unequivocal in their evidence.

Marine bands, although there are not many of them, allow much further refinement for they have the special feature that any one band is both very thin (a few feet, a few inches, sometimes scarcely more than a film) and widely persistent; and as they are intercalations in the midst of thick non-marine sediments, each one was formed virtually at a geological moment and thus represents a single time-plane over the whole coalfield (and beyond). Some fossil species found in them are relatively long-ranging, and may recur in successive bands; but the more important bands are mainly characterized by peculiar species or associations of species and carry their individuality wherever they are found. The principal bands are:

5. Upper Cwmgorse[1] Marine Band: with *Anthracoceras cambriense, Politoceras [Homoceratoides] kitchini, Peripetoceras [Cyclonautilus] dubium, Huanghoceras postcostatum, Dunbarella macgregori, Myalina compressa,* nuculids, productids (with *Dictyoclostus craigmarkensis*), spirifers, and ostracods. It is equivalent to the Top Marine Band of the Pennine coalfields and is the highest marine band known in the British Coal Measures.
4. Cefn Coed Marine Band: with *Anthracoceras aegiranum, Gastrioceras* aff. *globulosum, Politoceras politum [Homoceratoides jacksoni],* brachiopods (including *Rugosochonetes skipseyi, Tornquistia diminuta*), bivalves,

---

[1] The spelling 'Cwmgorse', used in the literature, is anglicized: the root is the Welsh '-gors' ('bog' or 'swamp'), not the English '-gorse' (a yellow-flowered spiny plant).

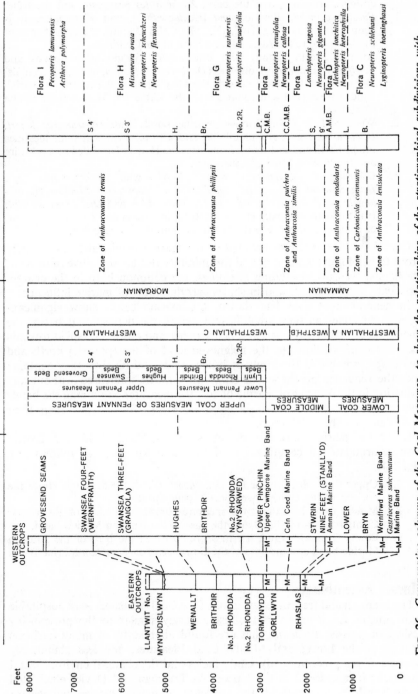

Fig. 26. *Comparative columns of the Coal Measures, to show the relationships of the stratigraphical subdivisions with each other and with the distribution of coal seams and marine beds*

crinoids, trilobites, and the coral *Zaphrentites postumus*, and the fishes *Edestus pringlei*, *Rhabdoderma* [*Coelacanthus*], and *Platysomus*. It is equivalent to the Mansfield, Gin Mine, and Dukinfield marine bands of England, and to the Aegir Marine Band of the Continent.

3. Amman Marine Band: with brachiopods (including productids and *Spirifer pennystonensis*), bivalves (*Dunbarella*, *Myalina*, nuculids), crinoids, and sponges. It correlates with the Clay Cross and Pennystone marine bands of England and equates with the Katharina Marine Band of the Continent.

2. Wernffrwd or Cefn Cribbwr Marine Band: with *Gastrioceras listeri*, *G. circumnodosum*, and nautiloids, and with productids at some localities on the north and east crops. It matches the Halifax Hard Bed of Yorkshire and the Bullion Mine of Lancashire.

1. *Gastrioceras subcrenatum* Marine Band: with *Anthracoceras spp.*, *Homoceratoides divaricatus*, *Dunbarella* cf. *papyracea*, *Myalina* cf. *sublamellosa*, *Polidevcia acuta*, *Lingula mytilloides*, and *Orthotetes cantrilli*. It is the conventionally chosen horizon, in Britain and on the Continent, of demarcation between the Namurian and the Westphalian.

Conodonts are widely distributed in the marine bands, and may shortly prove to be of great zonal use; and foraminifers also are definitive of some of the bands. 'Worms,' including *Planolites*, are common both in the characteristically marine bands and in shales carrying hints of marine influence.

In general, as Calver has shown, the marine bands show significant changes in facies as they are followed across the depositional basin, the fossil assemblage in any one band being richer and more diversified towards the near-shore shallows in the neighbourhood of the (present) north and east crops, notably in the kinds of brachiopods and bivalves they contain; and, the sequence on the north crop being much condensed without any reduction in the number of marine bands, the measures containing the bands give the impression there of being more richly fossiliferous than their equivalents on the south crop.

Using the marine bands as datum horizons, Stubblefield and Trotter have reformulated a classification of the Coal Measures as follows (*see* Fig. 26):

3. Upper Coal Measures or Pennant Measures: all the measures above the horizon of the Upper Cwmgorse Marine Band.

2. Middle Coal Measures: the measures between the top of the Upper Cwmgorse Marine Band and the base of the Amman Marine Band.

1. Lower Coal Measures: the measures between the base of the Amman Marine Band and the base of the *Gastrioceras subcrenatum* Marine Band.

Mussel Assemblages

The brackish and freshwater swamps of the Coal Measures were hospitable environments for a group of molluscs, closely similar to the present-day freshwater mussel *Anodonta*, that are found abundantly at many horizons throughout the Lower and Middle Coal Measures, and less abundantly (because of the predominantly sandy facies and the changed environmental conditions it represents) in the Upper. As Trueman and Davies elucidated in the South Wales coalfield, the sequence of mussel species and genera shows systematic change—an aspect of their evolution—that can be utilized to divide the measures into zones of very wide application. The mussels are referred to the genera *Carbonicola, Anthracosia, Anthracosphaerium, Anthraconaia, Naiadites, Curvirimula,* and *Anthraconauta* (*see* Fig. 27). Of these, only *Anthraconauta* is found in any numbers in the Upper Coal Measures, to which it is virtually restricted. All the other genera, except for a few occurrences of *Anthraconaia,* are confined to the Lower and Middle Coal Measures (in which all the marine bands also are to be found). Trueman therefore classed the two major rock groups, in reflection of this fossil distribution, into the Ammanian stage below and the Morganian

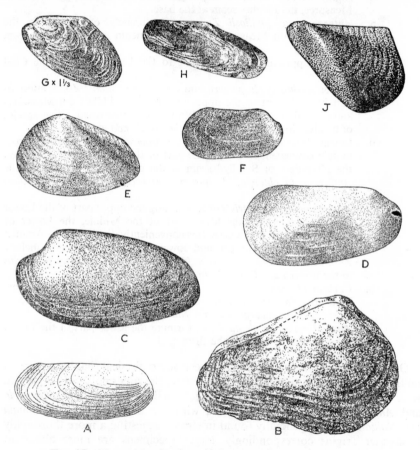

FIG. 27. *Non-marine bivalves of the Upper Carboniferous rocks*

(All figures natural size except **G.**)

A. *Anthraconaia lenisulcata* (Trueman); **B.** *Carbonicola pseudorobusta* Trueman; **C.** *Carbonicola communis* Davies and Trueman; **D.** *Anthraconaia modiolaris* (J. de C. Sowerby); **E.** *Anthracosia simulans* Trueman and Weir; **F.** *Anthraconaia pulchra* (Hind); **G.** *Anthraconauta phillipsii* (Williamson); **H.** *Anthraconauta tenuis* (Davies and Trueman); **J.** *Naiadites modiolaris* (J. de C. Sowerby).

B. is from J. Weir and D. Leitch, 1936, The zonal distribution of the non-marine lamellibranchs in the Coal Measures of Scotland, *Trans. Roy. Soc. Edin.* **50;** C and J are from *British Palaeozoic Fossils* published by the British Museum (Natural History). Permission to reproduce is acknowledged.

stage above. Within the Ammanian, *Carbonicola* is abundant only in the Lower Coal Measures, *Anthracosia* only in the lower part of the Middle Coal Measures (beneath the horizon of the Cefn Coed Marine Band), *Anthracosphaerium* in the upper part of the Lower Coal Measures and the lower part of the Middle Coal Measures, and *Naiadites* throughout the stage. In terms of the mussel sequence Trueman created the following zones (*see* Fig. 26):

2. Morganian (approximately equivalent to the Upper Coal Measures):
    b. *Anthraconauta tenuis* Zone, equivalent to the Upper Pennant Measures, the Hughes seam at the base.
    a. *Anthraconauta phillipsii* Zone, approximately equivalent to the Lower Pennant Measures, the Lower Pinchin or Tormynydd seam at the base.
1. Ammanian (approximately equivalent to the Lower and Middle Coal Measures):
    e. Upper *Anthraconaia pulchra* and *Anthracosia similis* Zone, approximately equivalent to the upper part of the Middle Coal Measures, and defined at its base by the Cefn Coed Marine Band. Other species of bivalves in the zone include *Anthraconaia adamsi* and *A. hindi*.
    d. Lower *Anthraconaia pulchra* and *Anthracosia similis* Zone, approximately equivalent to the middle part of the Middle Coal Measures, the Nine-Feet or Stanllyd seam at the base. Other species include *Anthraconaia oblonga*, *A. librata*, *Anthracosia atra*, and *Naiadites obliquus*.
    c. *Anthraconaia modiolaris* Zone, spanning the upper part of the Lower Coal Measures and the lower part of the Middle, the Lower or Seven-Feet seam at the base. It is conveniently divided at the Amman Marine Band into upper and lower parts. Other species include *Anthraconaia salteri*, *Anthracosia aquilina*, and *Naiadites quadratus* in the upper part, *Carbonicola cristagalli* in the lower.
    b. *Carbonicola communis* Zone, in the middle part of the Lower Coal Measures, the Bryn (Garw) seam at the base. Other species include *Carbonicola pseudorobusta* and *Naiadites flexuosus*.
    a. *Anthraconaia lenisulcata* Zone, forming the lower part of the Lower Coal Measures beneath the Bryn coal.

On a minor scale there seem to have been some physical (habitat) controls on the local distribution of the different genera, the zones not being expressed with uniformity everywhere. Large thick-shelled species of *Carbonicola* and *Anthracosia* are rarely associated with the thin-shelled *Anthraconaia* and *Naiadites*, being generally found in strata suggesting a more thoroughly freshwater origin: correspondingly, larger specimens are more abundant in the measures of the north crop (nearer the margins of the contemporary sedimentary basin) than of the south crop.

Other Fossils

Invertebrate fossils other than mussels are relatively rare and are not so useful for zonal purposes. Dix and Pringle, in a comprehensive description of the xiphosurs, showed that the broad twofold division of the rocks is maintained in the distribution of *Belinurus* and *Euproops*, the former being not uncommon in the Lower and Middle Coal Measures, the latter being

(A.9766)

A. Cwmparc in the Rhondda

B. Pennant escarpment at Craig-y-Llyn near Hirwaun

(A.4905)

Typical fossil plants of the Coal Measures

1. Part of the stem of a clubmoss; *Lepidodendron sp.* × ½. **2.** Part of the frond of a seed-fern; *Mariopter nervosa* (Brongniart). × 1. **3.** Leaves of a horsetail; *Annularia radiata* Brongniart. × 1. **4.** Part of the frond of tree-fern; *Asterotheca miltoni* (Artis). × ¾. **5.** Part of the leaf of a fern-like plant; *Sphenopteris* cf. *sancti-felic* (Stur). × 1. **6.** Pith-cast of the stem of a horsetail; *Calamites undulatus* Sternberg. × ⅔. **7.** Part of the frond of seed-fern; *Neuropteris scheuchzeri* Hoffman. × 1.

found mainly in the Upper. Other arthropods include the crustaceans *Euestheria, Leaia,* and *Carbonita,* common at certain horizons; and there are rarer occurrences of blattoid and other insects (*Archimylacris, Phylomylacris, Phyloblatta, Lithosialis, Boltoniella*).

Plant incrustations are the commonest fossils in the shales, and petrifactions are often found in the sandstones. Most of the plants belong to families that have long been extinct. The principal groups are the horsetails (Equisetales, such as *Calamites* and *Annularia*): the club-mosses (Lycopodiales, such as *Lepidodendron* and *Sigillaria*): the pteridosperms (plants with leaves similar to those of true ferns but differing from ferns in bearing seeds—*Neuropteris, Alethopteris,* and *Mariopteris*): the true ferns including *Asterotheca*; and the gymnosperms (related to the living conifers, and represented by *Cordaites*) (*See* Pl. VIII).

In a clarification of the detailed plant sequence, Dix has proposed seven major assemblage zones ('floras') in the South Wales measures and related them to the units of the Continental Westphalian stage. Recent work, notably by Sullivan and Butterworth, on the distribution of spores and pollen in the measures shows that palynological zones may be used to supplement the macrofloral zones.

In European correlation with South Wales, Westphalian A is approximately equivalent to the Lower Coal Measures, Westphalian B to the lower part of the Middle up to the horizon of the Cefn Coed Marine Band, Westphalian C to the upper part of the Middle plus the Lower Pennant Measures of the Upper Coal Measures, and Westphalian D to the Upper Pennant.

Although there is close conformity between the zonal schemes established on the three bases of the goniatite bands, the mussels, and the plants, it is not precise, the boundaries between the mussel and plant zones generally being arbitrarily placed at coal-seams (and not marine bands), and the upper two mussel zones spanning the same measures as the upper three plant zones (*see* Fig. 26).

Lower and Middle Coal Measures

The shales of the lower part of the Coal Measures crop out in major development at the foot of the 'Pennant' escarpment around the coalfield. They also emerge in the core of the Maesteg anticline, and appear in narrow inliers in the floors of some of the deeply entrenched valleys of the Rhondda (Figs. 28, 29). They consist for the most part of a monotonous series of grey, blue, and black pyritous and micaceous shales often imperfectly laminated. Some of the beds are highly carbonaceous and pass into a cannel shale or a cannel coal. Usually there is only a minor development of sandstones; but over much of the coalfield cockshot rock, a hard siliceous sandstone or grit, forms characteristic beds between the Red and Stwrin coals or their equivalents; and locally other impersistent sandstones may reach considerable thicknesses. The measures contain the greatest number, the thickest, and the most important coal seams in South Wales.

A feature of the measures, particularly of the lower part, is the abundance of iron ore occurring as a penecontemporaneous segregation of argillaceous chalybite (clay ironstone) in nodules or bands ('pins') arranged parallel with

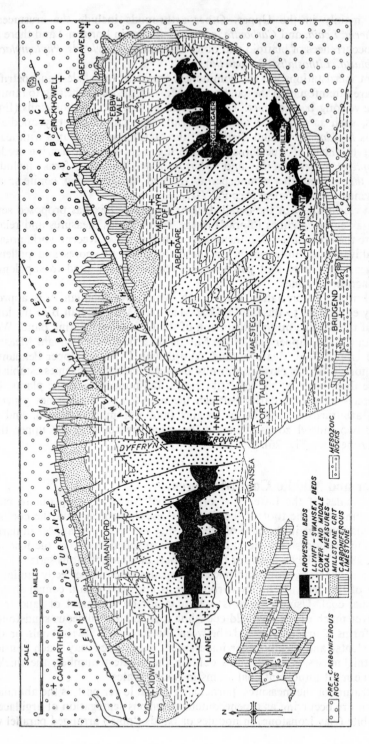

FIG. 28. Map of the South Wales coalfield

the stratification. At one time the ore formed the principal source of industrial iron in South Wales, but none is now worked. There is an increase in quantity but a decrease in quality of the ore as the ironstone bands are followed westwards and south-westwards.

At maximum the thickness of the Lower and Middle Coal Measures reaches some 3000 ft in the Swansea district, but it diminishes as the group is

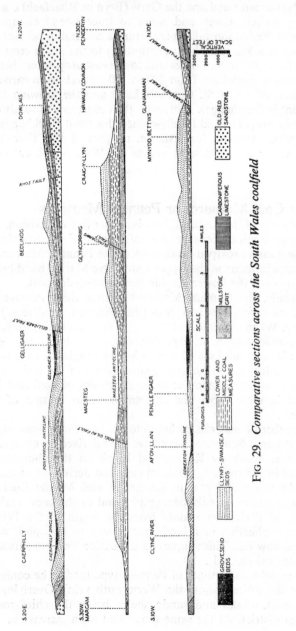

FIG. 29. *Comparative sections across the South Wales coalfield*

traced towards the north-east and east. In the Maesteg district it is about 2100 ft; near Merthyr Tydfil and in the south-east near Cardiff, 1400 ft; in the Pontypool district and the outcrops about Abertillery, slightly less than 800 ft; and southwards from Pontypool towards Risca it may be reduced to insignificance by repeated non-sequence.

Marine bands are recurrent in the measures (the *lenisulcata* Zone) between the *subcrenatum* Band and the Garw (Bryn or Rhasfach) coal, and have been used by Leitch, Owen and Jones to trace lateral changes in relation to thickness. Reduction takes place from a maximum of over 1000 ft between Swansea and Margam, both northwards to the north crop where the thickness falls to 300 ft, and eastwards to the east crop where between Brynmawr and Risca it diminishes from 200 to 100 ft; and, as in earlier sediments, the combined effects of St. George's Land and the growing Usk anticline are apparent in the changes. Moreover, the occurrence of sandstones is shown to be highly irregular, and locally-named 'Farewell Rock' appears at a number of horizons, so that, for instance, no part of the 'Farewell Rock' of the Neath valley is of the same age as the 'Farewell Rock' of the Brynmawr district.

Upper Coal Measures or Pennant Measures

Over most of the coalfield there is a clear line of division, both lithological and topographical, between the soft shales of the Middle Coal Measures and the massive scarped sandstones of the Pennant Measures (*see* Pl. VIIB), the precise junction at the Upper Cwmgorse Marine Band being a little below the horizon of the Lower Pinchin (or Tormynydd) coal.

Typically, the Pennant Measures include thick massive feldspathic and micaceous sandstones and grits (the 'Pennant Sandstone') much used for building. When unweathered the sandstones are bluish grey, but they rapidly become rusty-brown on exposure. Many bands are strongly cross-bedded and indicate conditions of deposition in proximity to a land-mass rapidly eroded by fast and powerful streams. Some of the beds are coarse-grained and may be conglomeratic: derived pebbles of coal and ironstone incorporated in them prove the contemporaneous erosion of already lithified measures.

A number of minerals common to the two formations suggests that much of the Pennant Sandstone may have been derived by erosion of the Old Red Sandstone. Work by Kelling and Bluck on the abundant washouts and channels in the rocks suggests an expected derivation of the detritus mostly from the north and east in the Lower and Middle Coal Measures; but, contrary to inferred palaeogeography and to thickness variations, a wholly unexpected derivation mainly from the south in the Pennant Measures, with its implication of a land-mass nearby to the south where the Bristol Channel now lies. This supports a suggestion first put forward by Heard on mineralogical grounds.

The massive sandstones of Pennant type tend to be confined to the lower part of the series (below the Wernffraith vein). Overlying beds are more argillaceous, shales and sandy shales being the chief rock types in the Swansea district. At the same time there is an increase in the number and quality of the coal seams present, and, with the Wernffraith seam, several higher coals have been extensively worked between the valleys of the Neath and the Gwendraeth.

On the basis of the changes occurring at the horizon of the Hughes vein, and using some of the more prominent and persistent coal seams as convenient markers, Woodland and his colleagues have recently reorganized and systematized the classification of the Upper Coal Measures, hitherto chaotic, in two major groups and six subgroups (*see* Fig. 26):

2. Upper Pennant Measures: the measures upwards from the Hughes seam, including the beds, the highest preserved in South Wales, formerly called the Upper Coal Series:
   c. Grovesend Beds, the Wernffraith or Mynyddislwyn seam at the base: a group relatively rich in workable seams.
   b. Swansea Beds, the Swansea Three-Feet or Graigola seam at the base.
   a. Hughes Beds, the Hughes or Wenallt seam at the base.
1. Lower Pennant Measures: the measures containing the main beds of typical massive Pennant Sandstone:
   c. Brithdir Beds, the Brithdir seam at the base.
   b. Rhondda Beds, the No. 2 Rhondda or Ynysarwed seam at the base.
   a. The Llynfi Beds, their base at the top of the Upper Cwmgorse Marine Band: the formation including the massive Pennant-like (but diachronous) Llynfi Rock.

Beds of red clay and mudstone are found in the Lower Pennant Measures especially in the eastern outcrops. They have been called the Deri Beds by Cox and Howell, and resemble in appearance the Etruria Marls of Staffordshire and the Ruabon Marls of North Wales. They may owe their origin, as Downing and Squirrell have argued, to the deposition of red-soil detritus transported from an upland source area, the colour possibly intensified by tropical weathering. Similar red clays at higher horizons in the Swansea and Grovesend Beds farther west are regarded by Archer also as being the product of penecontemporaneous weathering. The highest members of the Grovesend Beds in the Caerphilly basin, formerly called the Supra-Llantwit Measures, consist of red sandstones and marls and purple shales devoid of coal-seams: they indicate the onset of the arid climate that contributed to the change in conditions from Coal Measures to New Red Sandstone.

Because of post-Carboniferous erosion the complete original thickness of the Grovesend Beds is not preserved in South Wales, and the measurements of residual thicknesses are inconclusive. Nevertheless, the red beds occurring in the highest measures give some (though uncertain) indication of considerable thickening towards the south-west and west. Thus in the Gowerton syncline over 800 ft of normal grey coal-bearing strata form the greater part of some 1000 ft preserved, whereas in the Gelligaer and Llantwit–Caerphilly synclines the red beds appear only 300 ft above the Mynyddislwyn seam.

The Pennant Measures exceed 5000, perhaps 6000, ft in thickness in the western part of the coalfield about Swansea and Gorseinon. They are reduced to about 3500 ft in the Dyffryn trough near Neath, and to about 2000 ft (but in discontinuous sequence) in the Llantwit syncline. At the same time, there is a great reduction eastwards in the development of massive sandstones: the Pennant Sandstone as a diachronous formation has its base in the Rhondda Beds of the western outcrops (where it is underlain by the equally massive Llynfi Rock), but in the Brithdir Beds in eastern outcrops, where the underlying measures have only thin and unimpressive sandstones.

The Pembrokeshire Coalfield

In the Pembrokeshire coalfield the intense folding, crumpling, and faulting imposed by post-Westphalian movements so disturbed the stratal succession that it is difficult to recognize the divisions seen in the ground farther east; but Jenkins has shown that the sequence is broadly similar in order of thickness and in lithology to the measures from the base to the Swansea Beds of the main coalfield. The principal workable seams of the Lower Coal Measures match precisely the seams of the Gwendraeth anthracite belt, and are similarly anthracitic. The *Gastrioceras subcrenatum*, Amman, and Cefn Coed (Picton Point) marine bands are present; and Pennant-like sandstones (including the Rickets Head Sandstone) are well represented in the upper part of the sequence.

Contemporary Earth-movements

It is probable that during the formation of the Coal Measures, as of the Millstone Grit, the margins of the basin of deposition lay broadly to the north of the present outcrops. Nevertheless the maximum thinning takes place in South Wales not directly towards the north but towards the north-east (*see* Fig. 30), consistently in all three divisions, from a total of about 8000 to 10 000 ft near Swansea to less than 2000 ft around Pontypool and Abertillery; and even in the Caerphilly basin, on the same latitude as Swansea, the thickness is only about 3000 ft. It follows, therefore, that there was deflexion of the east–west 'shore-line' in the neighbourhood of the present eastern margin of the coalfield, and the strong presumption is that the Usk–Cardiff anticline was undergoing repeated uplift throughout the Westphalian period. An extreme reduction takes place along the east crop north of Risca, where the Lower and Middle Coal Measures together are reduced to less than 100 ft, perhaps to nil, and the Llynfi and Rhondda Beds together to about 190 ft. In the explanation offered by Moore and Blundell the great attenuation is a product of uplift and major overstep by the Pennant Measures, and of a complementary elimination by overlap of the Llynfi Beds, that matches the sub-Pennant unconformity in the Forest of Dean only 10 miles away. Squirrell and Downing have offered a variant interpretation in attributing the condensed sequence to slow subsidence with repeated minor breaks but without major sub-Pennant transgression. In either case, the contemporary Usk anticline was arched, or relatively remained a positive swell, in cumulative amplitude exceeding 2000 ft in a distance of little more than 8 miles.

Analogous signs of eastward overstep lie beneath the Grovesend Beds as the rocks are followed on to the northern flank of the Caerphilly basin, where the measures including the Mynyddislwyn seam descend on to horizons not far above the local equivalent of the Bettws (Wenallt or Hughes) seam, almost the whole of the Swansea Beds and the Hughes Beds being transgressed. Present limits of outcrop are distant some miles from the axis of the Usk–Cardiff anticline, and the magnitude of the unconformity over the anticlinal crest at the time of sedimentation cannot be assessed; but the scale of overstep is comparable with the condensation of the sequence below the Pennant Measures, and points to continued pulses of anticlinal uplift in later Westphalian times, perhaps continuing south of the coalfield along the Cardiff–Cowbridge axis. (*See* Fig. 30.)

Other evidence of relatively minor earth-movement is provided by thickness variations in the measures between the coal seams. Though these may be due partly to differential settling and compacting and to the lenticular form of sandy beds, Davies and Cox showed that the initiation of some of the north-and-south faults (at least as local sagging) must have taken place during the time of formation of the Lower Coal Measures; and Owen has provided evidence that the caledonoid Neath and Tawe disturbances are also in part of Carboniferous age.

Coals of South Wales

Although forming an insignificant proportion (less than 2 per cent) of the total thickness, the coal seams are economically the most important beds of the Coal Measures. The coals vary considerably in their properties and in their potential uses and fall into three main types—bituminous coals, steam coals, and anthracites. Though these kinds are readily distinguished in typical samples, they grade into one another and precise definition rests on fine chemical differences, as Seyler, Pollard, Radley, Strahan, Hicks, and Adams have shown. The bituminous coals are comparatively soft and friable; they yield a high proportion (20 to 40 per cent) of volatile matter and are good house, gas, and coking coals whose carbon content ranges from 84 to 91 per cent. Anthracite, on the other hand, is a hard stone coal of metallic lustre yielding a low proportion (3 to 8 per cent) of volatile matter and having a low hydrogen-content. It burns at high temperatures without yellow flame or smoke, and is unsuitable for the manufacture of coke. The content of carbon is high (more than 93 per cent), that of ash is often low. Steam coals are intermediate in composition and properties between the bituminous and the anthracitic types; they grade on the one hand into semi-bituminous and on the other into semi-anthracite coals.

There is a twofold systematic lithological variation in the development of the different kinds of coal in South Wales that shows features of unusual interest. On the one hand, in accordance with Hilt's 'law', the lower seams in the sequence at any one locality tend to be more anthracitic—to be of higher rank—than the higher seams: anthracites are rare in the Upper Coal Measures. On the other hand, any one coal-seam tends to become progressively more anthracitic as it is followed towards the north, north-west, and west: thus the bituminous coals are mainly found along the southern and eastern outcrops, the steam coals in the central part of the coalfield between the Taff and the Neath (particularly in the Rhondda) valleys, and the anthracites along the north crop westwards from the Neath valley, especially in the Gwendraeth valley and in Pembrokeshire. (*See* Fig. 30.)

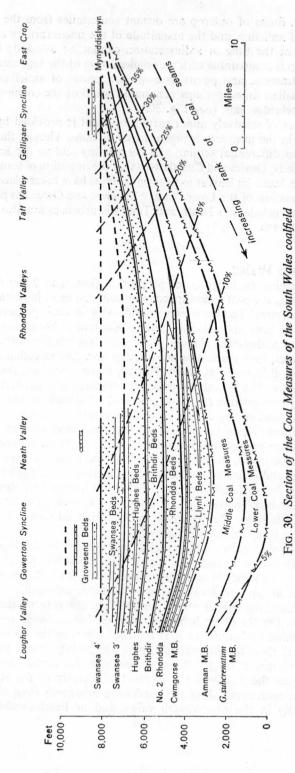

FIG. 30. *Section of the Coal Measures of the South Wales coalfield*

The reconstruction is generalized and runs approximately west to east. The variable development of sandstones ('Llynfi Rock', 'Pennant Sandstone') is diagrammatically indicated. The progressive stages of change from high-volatile bituminous coals in the east to anthracite in the west is indicated by the generalized isovols.

The reasons for the regional changes are not certainly known. They do not appear to be due to differences in the plant composition of the initial peats, for the association of plant debris in anthracites is much as it is in low-rank coals. Trotter argued that devolatilization, and the corresponding increase of rank from gas coals to anthracite, were products of shearing stress correlated in South Wales with the thrust plane that comes to out-crop as the Careg Cennen disturbance on the north crop and that sup-posedly sinks to depths southwards underneath the coalfield: the rank of a coal-seam is then a function of its distance from the thrust plane. In con-trast, Jones and Wellman have invoked a process of load metamorphism that is an aspect of Hilt's 'law', the weight of overlying sediment being the initial load, and rank being then a function of pressure and temperature at depth. Such an explanation, applied to the present anthracite field, implies that where the relatively thin representatives of the Lower and Middle Coal Measures of the Gwendraeth belt now lie there was formerly a very thick overburden—perhaps (including post-Carboniferous rocks) of the order of 20 000 ft. It also implies that the depositional basin carried its thickest accumulation of sediments in the region of the Gowerton and Llanelli synclines, where semi-anthracites occur as high in the sequence as the seams of the Swansea Beds.

References

ADAMS, H. F. 1956. Seam structure and thickness in the South Wales coalfield. *Proc. S. Wales Inst. Eng.*, **71**, 96–106.
—— 1967. The seams of the South Wales coalfield. *Mon. Inst. Min. Eng.*
ARCHER, A. A. 1965. Red beds in the Upper Coal Measures of the western part of the South Wales coalfield. *Bull. Geol. Surv. Gt Brit.* No. 23, 57–64.
BLOXAM, T. W. 1964. Uranium, thorium, potassium, and carbon in some black shales from the South Wales coalfield. *Geochim. & Cosmochim. Acta*, **28**, 1177–85.
BLUCK, B. J. and KELLING, G. 1963. Channels in the Upper Carboniferous Coal Measures of South Wales. *Sedimentology*, **2**, 29–53.
BLUNDELL, C. R. K. 1952. The succession and structure of the north-eastern area of the South Wales coalfield. *Quart. J. Geol. Soc.*, **107**, 307–33.
D.S.I.R. 1944. A description of an isovol map of the South Wales coalfield. *Rep. Fuel. Res. Bd.*, **56**.
DAVIES, D. 1921. The Ecology of the Westphalian and the lower part of the Staffordian Series of Clydach Vale and Gilfach Goch. *Quart. J. Geol. Soc.*, **77**, 30–74.
—— 1929. Correlation and palaeontology of the Coal Measures of east Glamorganshire. *Phil. Trans. Roy. Soc.*, (B), **217**, 91–154.
DAVIES, D. F., DIX, E. and TRUEMAN, A. E. 1928. Boreholes in Cwmgorse valley. *Proc. S. Wales Inst. Eng.*, **44**, 37–136.
DAVIES, J. H. and TRUEMAN, A. E. 1923. The correlation of the Coal Measures in the western portion of the South Wales Coalfield. *Proc. S. Wales Inst. Eng.*, **39**, 367–91.
—— —— 1927. A revision of the non-marine lamellibranchs of the Coal Measures and a discussion of their zonal sequence. *Quart. J. Geol. Soc.*, **83**, 210–59.
DAVIES, R. 1922. Some effects of intra-Coal Measure movements: a further study. *Proc. S. Wales Inst. Eng.*, **38**, 611–21.

DAVIES, R. and COX, A. H. 1922. On thickness variations in the Lower Coal Measures of east Glamorganshire and Monmouthshire. *Proc. S. Wales Inst. Eng.*, **39**, 41–62.

DIX, E. 1928. The Coal Measures of the Gwendraeth valley and adjoining areas. *Proc. S. Wales Inst. Eng.*, **44**, 423–510.

—— 1934. The sequence of floras in the Upper Carboniferous with special reference to South Wales. *Trans. Roy. Soc. Edin.*, **57**, 789–821.

—— and PRINGLE, J. 1929. On the fossil Xiphosura from the South Wales coalfield with a note on the myriapod *Euphoberia*. *Sum. Prog. Geol. Surv.* for 1928, Pt 2, 90–114.

—— —— 1930. Some Coal Measure arthropods from the South Wales coalfield. *Ann. Mag. Nat. Hist.*, (10), **6**, 136–44.

—— and TRUEMAN, A. E. 1924. The Coal Measures of north Gower. *Proc. S. Wales Inst. Eng.*, **40**, 353–83.

—— —— 1928. Marine horizons in the Coal Measures of South Wales. *Geol. Mag.*, **65**, 356–63.

—— —— 1928. Some non-marine lamellibranchs from the upper part of the Coal Measures. *Quart. J. Geol. Soc.*, **87**, 180–211.

DOWNING, R. A. and SQUIRRELL, H. C. 1965. On the red and green beds in the Upper Coal Measures of the eastern part of the South Wales coalfield. *Bull. Geol. Surv. Gt Brit.*, No. 23, 45–56.

EVANS, W. H. and SIMPSON, B. 1934. The Coal Measures of the Maesteg district. *Proc. S. Wales Inst. Eng.*, **49**, 447–75.

HEARD, A. 1922. The petrology of the Pennant Series. *Geol. Mag.*, **59**, 83–92.

HOWELL, A. 1927. The correlation of the coals of the Lower Coal Series in the east Glamorgan and Monmouthshire coalfield. *Proc. S. Wales Inst. Eng.*, **43**, 321–81.

—— and COX, A. H. 1924. On a group of red measures or 'coloured' strata in the Glamorganshire and Monmouthshire coalfield. *Proc. S. Wales Inst. Eng.*, **40**, 139–74.

JENKINS, T. B. H. 1962. The sequence and correlation of the Coal Measures of Pembrokeshire. *Quart. J. Geol. Soc.*, **118**, 65–101.

JONES, D. G. and SQUIRRELL, H. C. 1969. The discovery of the *Gastrioceras subcrenatum* Marine Band at Risca, Monmouthshire. *Bull. Geol. Surv. Gt Brit.*, No. 30, 65–70.

JONES, O. T. 1949. Hilt's Law and the volatile content of coal seams. *Geol. Mag.*, **86**, 303–12, 346–64.

—— 1951. The distribution of coal volatiles in the South Wales coalfield and its probable significance. *Quart. J. Geol. Soc.*, **107**, 51–83.

JONES, S. C. 1957. The northward attenuation of the coal seams in the Swansea district. *Trans. Inst. Min. Eng.*, **116**, 445–56.

JONES, S. H. 1934. The correlation of the coal seams of the country around Ammanford. *Proc. S. Wales Inst. Eng.*, **49**, 409–46.

—— 1935. The Lower Coal Series of north-western Gower. *Proc. S. Wales Inst. Eng.*, **50**, 317–81.

JORDAN, H. K. 1903. Notes on the south trough of the east Glamorgan coalfield. *Proc. S. Wales Inst. Eng.*, **23**, 131–56.

—— 1910. The South Wales coalfield: sections and notes. *Proc. S. Wales Inst. Eng.*, **26**, 172–254.

KELLING, G. 1964. Sediment transport in part of the Lower Pennant Measures of South Wales. In *Developments in Sedimentology*, **1**, 177–84.

KUENEN, P. H. 1948. Slumping in the Carboniferous rocks of Pembrokeshire. *Quart. J. Geol. Soc.*, **104**, 365–80.

LEITCH, D., OWEN, T. R. and JONES, D. G. 1958. The basal Coal Measures of the South Wales coalfield from Llandebie to Brynmawr. *Quart. J. Geol. Soc.*, **113**, 461–86.

MOORE, L. R. 1945. The geological sequence of the South Wales coalfield: the 'south crop' and Caerphilly basin, and its correlation with the Taff valley sequence. *Proc. S. Wales Inst. Eng.*, **60**, 141–227.

—— 1947. The sequence and structure of the southern portion of the east crop of the South Wales coalfield. *Quart. J. Geol. Soc.*, **103**, 261–300.

—— and BLUNDELL, C. R. K. 1952. Some effects of the Malvernian phase of earth-movements in the South Wales coalfield, a comparison with other coalfields in south Britain. *C. R. 3me Congr. Strat. Geol. Carb.*, **2**, 463–73.

—— and COX, A. H. 1943. The Coal Measure sequence in the Taff valley, Glamorgan, and its correlation with the Rhondda valley sequence. *Proc. S. Wales Inst. Eng.*, **59**, 189–286.

NORTH, F. J. 1931. Coal and the coalfields in Wales. 2nd Edition. Cardiff. (Nat. Mus. Wales.)

—— 1935. The fossils and the geological history of the South Wales Coal Measures. *Proc. S. Wales Inst. Eng.*, **51**, 271–300.

RAMSBOTTOM, W. H. C. 1952. The fauna of the Cefn Coed marine band in the Coal Measures at Aberbaiden, near Tondu, Glamorgan. *Bull. Geol. Surv. Gt Brit.*, No. 4, 8–30.

ROWLANDS, T. H. 1925. The upper part of the Pennant Series of the Swansea district. *Proc. S. Wales Inst. Eng.*, **41**, 257–98.

SEYLER, C. A. 1901. The chemical classification of coal. *Proc. S. Wales Inst. Eng.*, **21**, 483–526.

SMITH, A. H. V. and BUTTERWORTH, M. A. 1967. Miospores in the coal seams of the Carboniferous of Great Britain. *Spec. Publ. Palaeontological Assoc.*, **1**.

SQUIRRELL, H. C. and DOWNING, R. A. 1964. The attenuation of the Coal Measures in the south-east part of the South Wales coalfield. *Bull. Geol. Surv. Gt Brit.*, No. 21, 119–32.

STRAHAN, A. and POLLARD, W. 1915. The coals of South Wales. 2nd Edition. *Mem. Geol. Surv.*

STUART, A. 1924. The micropetrology of South Wales anthracite. *Geol. Mag.*, **61**, 360–6.

STUBBLEFIELD, C. J. and TROTTER, F. M. 1957. Divisions of the Coal Measures on the Geological Survey maps of England and Wales. *Bull. Geol. Surv. Gt Brit.*, No. 13, 1–5.

SULLIVAN, H. J. 1962. Distribution of miospores through coals and shales of the Coal Measures sequence exposed in Wernddu claypit, Caerphilly (South Wales). *Quart. J. Geol. Soc.*, **118**, 353–73.

TROTTER, F. M. 1948. The devolatilization of coal seams in South Wales. *Quart. J. Geol. Soc.*, **104**, 387–437.

—— 1950. The devolatilization equation for South Wales coals. *Geol. Mag.*, **87**, 196–208.

—— 1954. The genesis of high rank coals. *Proc. Yorks. Geol. Soc.*, **29**, 267–303.

TRUEMAN, A. E. 1933. A suggested correlation of the Coal Measures of England and Wales. *Proc. S. Wales Inst. Eng.*, **49**, 63–106.

—— 1934. The age of the highest Coal Measures in West Pembrokeshire. *Geol. Mag.*, **71**, 116–8.

—— 1941. The periods of coal formation represented in the British Coal Measures. *Geol. Mag.*, **78**, 71–6.

—— 1946. Stratigraphical problems in the Coal Measures of Europe and North America. *Quart. J. Geol. Soc.*, **102**, xlix-xciii.

TRUEMAN, A. E. and WARE, W. D. 1932. Additions to the fauna of the Coal Measures of South Wales. *Proc. S. Wales Inst. Eng.*, **48**, 67–92.

—— and WEIR, J. 1946–68. A monograph of British Carboniferous non-marine Lamellibranchia. *Palaeont. Soc.*

—— and OTHERS. 1954. *The coalfields of Great Britain*. London.

WARE, W. D. 1930. The geology of the Cefn Coed sinkings. *Proc. S. Wales Inst. Eng.*, **46**, 453–501.

WELLMAN, H. W. 1950. Depth of burial of South Wales coals. *Geol. Mag.*, **87**, 305–23.

WILSON, M. J. 1965. The origin and significance of the South Wales underclays. *J. Sed. Pet.*, **35**, 91–9.

—— 1965. The underclays of the South Wales coalfield east of the Vale of Neath. *Trans. Inst. Min. Eng.*, **124**, 389–404.

WOODLAND, A. W., ARCHER, A. A. and EVANS, W. B. 1957. Recent boreholes into the Lower Coal Measures below the Gellideg–Lower Pumpquart coal horizon in South Wales. *Bull. Geol. Surv. Gt Brit.*, No. 13, 39–60.

—— EVANS, W. B. and STEPHENS, J. V. 1957. Classification of the Coal Measures of South Wales with special reference to the Upper Coal Measures. *Bull. Geol. Surv. Gt Brit.*, No. 13, 6–13.

10. Palaeozoic Earth-movements

The major cycles of sedimentation revealed in the lithological contrasts between the rocks of the various systems are signs of the recurrence of orogenic movements in South Wales during Palaeozoic times. The end of the Lower Palaeozoic sequence at the base of the Old Red Sandstone, and the end of the Upper Palaeozoic at the local base of the New, offer a means of distinguishing between impulses of orogenesis that in Europe have been classified the earlier as Caledonian the later as Hercynian or Armorican. The radical transformation of Silurian seas to 'continental' Devonian deltas or of Westphalian swamps (after a long interval) to Triassic salt 'lakes' is nevertheless mainly represented in changes in rock facies, and is not itself proof of folding and faulting of a magnitude that would truly be called orogenic, which resides in strong unconformity and overstep. When tectonic relations and not lithological differences become the measure, it is found that Hercynian structures are not always easily distinguished in time or in place from Caledonian. (*See* Fig. 31.)

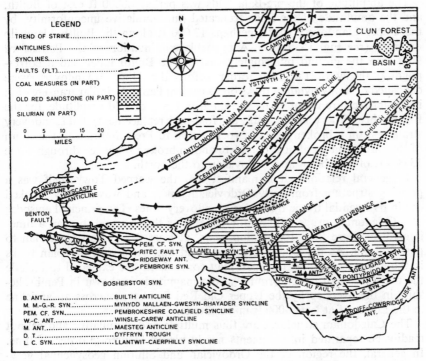

FIG. 31. *Map of the major structures in South Wales*

Caledonian Movements

A Caledonian 'period' of earth-movement is not readily defined in South Wales. Despite the marked changes in rock-type, the Old Red Sandstone rests upon Silurian rocks in apparent continuity, certainly without significant difference in dip, along much of the outcrop from Shropshire into mid-Wales; and only in Carmarthenshire and Pembrokeshire is discordance obvious in strong transgression by the younger system. The term Caledonian thus cannot well be simply equated with Siluro-Devonian.

On the other hand, major unconformities interrupt the sequence at various Lower Palaeozoic horizons. They are to be seen in the absence of the Tremadoc stage of the Cambrian and the local overstep of the Ordovician rocks on to low Cambrian horizons; in the transgression of Bala sediments on to various horizons of the Ordovician, probably on to a Cambrian and Pre-Cambrian floor, on the structural shelf east and south-east of the Towy anticline; in the even more widely extensive transgression of the Upper Llandovery sandstones of the Builth and Llandovery outcrops and of Pembrokeshire; and in the blanket of gently inclined Wenlock Shales that overlies the sharply folded rocks (down to Llanvirn) in mid-Wales and along the east flank of the Towy anticline (*see* Fig. 14, p. 39). Each of these hetero-chronous structures reaches a magnitude approaching or exceeding any to be recognized beneath the Old Red Sandstone—which, at its maximum overstep in Pembrokeshire, inherited tectonic relationships that were initiated before Llandovery times and were well established by mid-Silurian.

The amplitude of the pre-Bala folds was perhaps 8000 ft east of Builth, and the amplitude of the folds truncated (in cumulative unconformity) by the Upper Llandovery rocks, perhaps 12 000 ft along the Builth–Llandeilo outcrops (*see* Fig. 15, p. 42). A probable continuation of the Pontesford fault into the ground immediately east of the Builth inlier had a corresponding throw of the same order before the Upper Llandovery were deposited on Longmyndian rocks on its eastern flank. A similar transgression in Pembrokeshire is not easily measured in terms of fold amplitude, for the Benton thrust intervenes (possibly as a fold-limb replacement); but Llandovery strata rest on a broken series of rocks down to Pre-Cambrian to the south of the thrust, on an almost complete and unbroken sequence through some 13 000 ft of Cambrian and Ordovician to the north.

In general relations, it is manifest that the warped Towy shelf was a growing structure from early Ordovician times, and was already blocked out in its essential frame by Upper Llandovery times. The principal episodes of a Caledonian orogeny in South Wales may thus be said to span a time-interval of more than 100 million years, at least from latest Cambrian to post-Silurian. As the Lower Old Red Sandstone runs with Salopian rocks in the eastern outcrops, the span may conveniently be extended to include the mid-Devonian unconformity, whose magnitude is shown in Pembroke-shire in the overstep by the Upper Old Red Sandstone across an anticline in the Lower of at least 9000 ft in amplitude.

The Caledonian structures are thus multiple and cumulative and cannot readily be separated into elements neatly placed in precise order of time. In general, the region of the Ordovician and Silurian geosynclinal grey-wackes and shales north-west of the Towy anticline provides least evidence

of interrupted sedimentation, and in continued (if pulsed and undulatory)
subsidence appears to have been beyond the range of significant deformation
until the close of the Silurian period. In contrast, the neritic sediments of
the marginal shelves of the geosyncline east and south-east of the Towy
anticline show many breaks in the sequence and were repeatedly folded,
faulted, and eroded. The anticlinal belt thus delineates a major contrast in
tectonic style as well as in lithological facies between two Lower Palaeozoic
tracts in South Wales.

The main Caledonian structures are aligned in a north-east–south-west
(caledonoid) direction. They are impressively seen in the major folds of
mid-Wales (*see* Fig. 32). The Towy anticline is the chief: it may be followed
as a long inlier of Ordovician rocks from west Carmarthenshire deep into
Radnorshire, where it is closed by Upper Llandovery and Wenlock overstep.
On its south-east flank Lower Old Red Sandstone is as steeply dipping as
Silurian rocks, and a major pulse in its arching, in amplitude of several
thousand feet, probably took place in mid-Devonian times. It is flanked to
the east by the Builth anticline, scarcely less important in revealing a core of
basinal Ordovician rocks whose sub-surface extensions beneath transgressive
Silurian reappear north-eastwards in the Shelve region of Shropshire, and
must be supposed to continue indefinitely southwards into Breconshire.
To the south-east of the Builth anticline, at shallow depth beneath the
Silurian cover, there presumably lie Cambrian and Longmyndian rocks as in
Shropshire; and beyond the hidden continuation of the Longmynd horst,
the Church Stretton fault runs from Presteigne and Old Radnor into the
Brecon anticline, seen at the surface as a tongue of Ludlow shales in the
Old Red terrain of the back slopes of Mynydd Epynt.

The Teifi anticline is less acute and persistent than the Towy anticline.
It is well defined, in a blunt plunging core, by the north-eastward embayment
in the outcrop of the Silurian base between Lampeter and Cardigan; but as
it is followed into mid-Wales it is not readily identified in the wide Silurian
outcrops. Ordovician (Bala) rocks emerge in the Plynlimon culmination
and in smaller anticlinal cores to east and west, in folds that are multiple and
lack a dominant central axis. Correspondingly, the Central Wales syncline
lying between the Towy and Teifi anticlines is well displayed in the long nose
of Silurian rocks that runs south-westwards north of Carmarthen; but it is
complicated north-eastwards by the minor Cothi–Rhiwnant anticline on its
eastern flank, and it begins to lose its identity in the multiple folds about
Llanidloes, where its main element lies in the Tarannon basin.

Multiple periclines reduce the Silurian outcrops to fragments in north
Pembrokeshire, where the Pre-Cambrian, Cambrian, and Ordovician outcrops fall into a regional pattern of plunging folds, much faulted, whose
axes are as yet not readily traceable for long distances and cannot be
continued with certainty into the fold-axes in mid-Wales. A contrast in
tectonic style is perhaps to be attributed on the one hand to the broad
uniformity of the thick piles of sedimentary rocks in mid-Wales, and on the
other to the occurrences of highly variable but thick and massive igneous
rocks at various horizons in the sedimentary sequence in Pembrokeshire.

The rocks of the greywacke suite in the Towy and Teifi anticlines and the
Central Wales syncline appear in open folds on a small-scale map, with

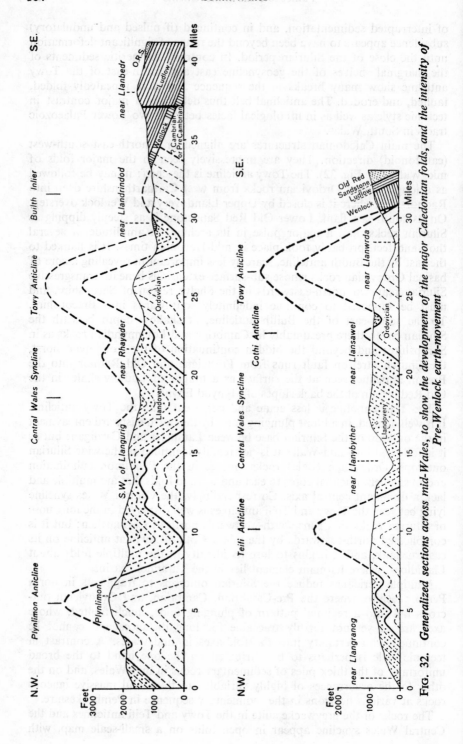

Fig. 32. *Generalized sections across mid-Wales, to show the development of the major Caledonian folds, and the intensity of Pre-Wenlock earth-movement*

notably wide outcrops of Silurian grits and shales forming most of the ground; but in detail a multitude of smaller-scale folds, some of them sharply acute, are imposed on the larger (*see* Pl. IXA), and locally a combination of compressive and shearing stresses has resulted in the finer-grained Ordovician and Silurian rocks being cleaved in discordance to the bedding: roofing slates were formerly extensively worked in Cardiganshire and Pembrokeshire, although they do not compare in quality with the Cambrian slates of Llanberis and Bethesda or the Ordovician slates of Blaenau Ffestiniog. The cleavage has been taken to imply a former thick cover of rocks removed by erosion in post-Llandovery times. In contrast, the shelf sediments south-east of the Towy anticline, not accumulated in a thick geosynclinal prism, are mostly free of cleavage.

The Caledonian folds of north Cardiganshire are usually acute and swing towards a northerly direction: they are then not wholly congruent with the style and the alignment of the structures on the south flank of the Merioneth dome, and show clear signs of disharmonic relationships between neighbouring formations or members offering differential resistance to stress. They are also cut obliquely by major faults trending north of east: of these, the Llyfnant, Camddwr, and Ystwyth faults demonstrably have a major component of dextral strike-slip in displacing earlier folds and faults, and represent a late-phase element in the structural pattern.

Hercynian Movements

After the mid-Devonian movements died away, there was little deformation of the magnitude of orogenic folding until late Carboniferous times. St. George's Land was broadly anticlinal in structure, and its continued growth affected thickness and sequence in the Carboniferous rocks, not least in the development of the several intra-Carboniferous unconformities; but although the amplitude of the (Upper Palaeozoic) anticlinal dome between North and South Wales is to be measured in thousands of feet, the dips of the tilted beds on its Carboniferous flanks were of the order of 5° or less, and even at the present day the dips along parts of the north crop of the South Wales coalfield are not much more.

Towards or shortly after the close of Westphalian times, on the other hand, accumulated stresses imposed strong folding and faulting on the Upper Palaeozoic rocks and gave them their present disposition. In the main, the dominant direction of stress was from the south, and the fold axes run more or less east-and-west: the structures are parallel to similar structures in Devon and Cornwall, and farther south in Brittany (Armorica), and fall into the Armorican arc of the Hercynian orogen of north-western Europe. Except in Pembrokeshire, they are thus discordant with the Caledonian structures of the Lower Palaeozoic rocks, and are conveniently distinguished as Hercynian because of the differences in age and in alignment. The folds are most acute in southern outcrops and decline in intensity northwards, perhaps because of dissipation in superficial structures of the orogenic pressures against the massif of St. George's Land. (*See* Fig. 31.)

The most outstanding Hercynian structure on the map is the syncline of the South Wales coalfield (*see* Fig. 28, p. 90). It runs from St. Bride's Bay

in west Pembrokeshire continuously through Carmarthenshire and
Glamorgan into Monmouthshire, and terminates along the east crop only
because of transection by the anomalous Usk anticline. In detail it is a
complex downfold consisting of a number of periclinal flexures arranged in
echelon (*see* Pl. IXB), each impersistent when traced along the axial strike.
Thus in the eastern part a central arch, the Pontypridd anticline, breaks the
simplicity of the elongate basin and separates two minor synclines—the
Gelligaer syncline and the Llantwit–Caerphilly syncline—in which Grovesend
Beds of the uppermost Coal Measures are preserved in isolated outliers.
Westwards the Pontypridd anticline dies away, the Maesteg anticline being
its phased analogue; but west of the River Neath the broad synclinal structure is relatively simple. Still farther west the Gowerton syncline and the
Llanelli syncline (in both of which Grovesend Beds are preserved) are
analogues of the Gelligaer and Caerphilly basins, as minor structures within
the coalfield. Still smaller and more localized folds are common, especially
in the less competent shales of the Lower and Middle Coal Measures—the
massive beds of the Pennant Sandstone being much more simply deformed.
In broad structure the Pembrokeshire branch of the coalfield has the form of
a simple syncline, but at most places the rocks are in detail much contorted
and crumpled, and the coals much broken (and therefore inferior in economic
quality).

To the south of the coalfield syncline the Upper Palaeozoic rocks are
arranged in elongate periclines showing strike replacement in echelon.
The intervening synclines are sufficiently deep to preserve Namurian
beds (but not Coal Measures) at outcrop. In the Vale of Glamorgan the
most prominent fold is the Cardiff–Cowbridge anticline, which reaches a
culmination at Cardiff where Salopian rocks emerge in the Rumney inlier:
its analogue farther west, not directly along the strike, is the Candleston
anticline. The similar Cefn Bryn anticline, with Old Red Sandstone in its
core, forms the 'backbone' of Gower. It is flanked or followed by sub-
sidiary or replacement anticlines, including the Langland, Llanmadoc
Hill, Oxwich Point and Worms Head anticlines, and by corresponding
synclines in which Namurian shales form pockets of soft rocks readily
etched by the sea to form Oxwich and Port Eynon Bays (*see* Pl. VIB).
Beyond Carmarthen Bay, three or four sub-parallel periclines dominate
southern Pembrokeshire. In the Winsle–Carew anticline and the more
southerly Freshwater East and Castlemartin–Corse anticlines the folding
is sufficiently acute to bring Lower Palaeozoic rocks to the surface; and in
the intervening Pembroke and Bullslaughter Bay synclines Namurian shales
are preserved. (*See* Fig. 33.)

The folds are broken by faults many of which appear to be complementary
to the folds. This is clearly so where as strike faults they act as fold-limb
replacements. Overthrusting is common, and in the weaker rocks, as in the
Pembrokeshire coalfield, may fracture the rocks into innumerable small
slices. The Caswell thrust (Pl. VA), breaking the multiple Langland anticline
in Gower, has a visible displacement of several hundred feet. Along the
Ritec thrust the Pembroke syncline was carried several thousand feet on to
the Sageston anticline; and along the Johnston thrust, a few miles farther
north, Pre-Cambrian rocks over-rode Coal Measures, in movement that

(*Cambridge Univ.*)

A. The cliffs of Pen-yr-Afr, near Cardigan

B. The cliffs at Little Haven, St. Bride's Bay

(A.921)

(*T. N. George*)

A. Unconformity at Cold Knap near Barry

B. The Lias cliffs between Llantwit Major and Aberthaw

(*T. N. George*)

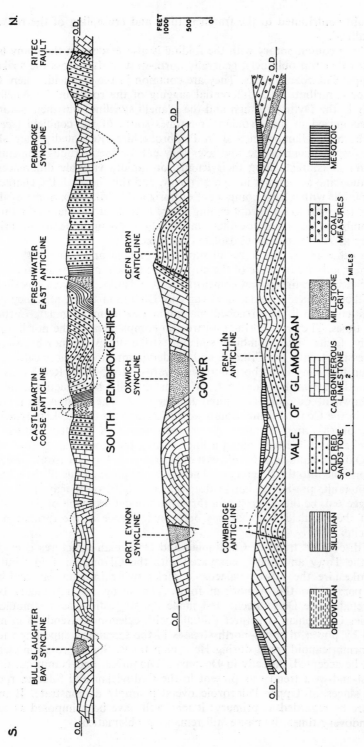

Fig. 33. *Sections across the Armorican folds of South Wales*

no doubt contributed to the fragmentation and crumpling of the rocks of the coalfield.

Faulting contemporary with the folding is also evident in the many large fractures that run obliquely, generally north-westwards, across the strike of the Upper Palaeozoic rocks. They are common in the coalfield, where they cause or contribute to a differential sagging of the rocks in the synclines, notably in the Dyffryn trough and the Llanelli syncline; but their nature is best recognized in the Avonian limestones south of the coalfield (*see* Pl. VIB), both in Glamorgan and in Pembrokeshire. Repeatedly they show lateral displacement of the fold axes they cut, or incongruous elements of structure in contact along their length, or rapidly variable differences in magnitude and sometimes in sign of throw, and thus have all the characters of tear faults with a large component of horizontal strike-slip. Some of them may be followed to an abrupt ending where they meet the plane of a thrust, the combined fractures together, manifestly contemporaneous in origin, defining a fault-bound block isolated on three sides.

Two other major structures disrupt the coalfield and cause interference with the regularity of strike of the north crop: these are the Neath and Tawe disturbances. Stratigraphical contrasts on their flanks, as Owen has shown, prove them to have originated as early as Dinantian times, and they grew intermittently until they reached maximum development in late-Carboniferous times. They run north-eastwards, in complement to the north-north-westerly faults, and combine sinistral strike-slip with north-westward compression seen in the Cribarth and Penderyn anticlines, whose caledonoid axes run for many miles both north-eastwards into the Old Red Sandstone and south-westwards into the Coal Measures. Similar structures, not so well shown at outcrop, are found farther west. The Llanon disturbance occurs in the Coal Measures south-west of Ammanford. The Llandyfaelog–Llangyndeyrn–Careg Cennen disturbance preserves faulted outliers of Carboniferous Limestone along a line running for more than twenty miles in the Old Red Sandstone terrain east of Carmarthen Bay (*see* frontispiece): it may continue into the Brecon anticline and so be an analogue of the Church Stretton fault, pre-Hercynian in deep-seated origin; or it may, as Trotter has suggested, be the outcrop of a thrust that underlies much of the western part of the coalfield and that has been an influence in the formation of anthracite.

The discordant trend of Caledonian and Hercynian structures is evident when the Towy anticline is compared with the Glamorgan folds; but in Pembrokeshire the Lower Palaeozoic rocks swing into an east-and-west strike parallel to the Armorican folds. The swing has repeatedly been interpreted, since De la Beche first made the suggestion, as a remoulding of earlier Caledonian structures (initially with caledonoid trend, as in mid-Wales) by powerful south–north stresses in the secondary superimposition of an armoricanoid trend during Hercynian times. Such a rotation clearly cannot be accepted. Already in Devonian, Dinantian, and Namurian times the east-and-west trend was present in the Ordovician and Silurian rocks, as the stages of Upper Palaeozoic overstep amply demonstrate. It must therefore be regarded as primary: it may well have been imposed as early as Llandovery times. Its cause still remains problematical.

Mineralization

Lead, zinc, copper, and iron ores are present as lodes and veins in many of the Palaeozoic rocks, particularly in the Llandovery rocks of mid-Wales and the Carboniferous Limestone of South Wales. The ore-bodies commonly follow faults and joints, and they may have been emplaced during a late-stage orogenic phase of hydrothermal injection—as Hercynian by-products in the Carboniferous rocks, less certainly as Caledonian by-products in the Llandovery. Formerly the exploitation of lead in Cardiganshire was highly profitable, but while there still may be considerable reserves at depth, mining is no longer economic. Small concentrations of gold in quartz lodes in the Llandovery rocks at Ogofau, north-west of Llandovery, are said to have been worked by the Romans.

The iron ores (mainly hematite) found as veins and large pods in the Carboniferous rocks are fissure or chamber fillings in part formed by metasomatic replacement of limestone country-rock. Still worked at the Llanharry mine, they are relatively richly developed along the south-east crop of the coalfield. Their origin is not certainly linked with the Hercynian movements, for while they may, on one theory, be derived by a hydrothermal leaching and oxidation of the siderite and chalybite of the bedded clay-ironstones of the Coal Measures, they may, on another, be derived iron-oxide concentrates of a Triassic terrain and thus post-Hercynian in age.

References

ANDERSON, J. G. C. and OWEN, T. R. 1967. *The structure of the British Isles*. London.

BASSETT, D. A. 1969. Some of the major structures of early Palaeozoic age in Wales and the Welsh borderland: an historical essay. *In* A. Wood (editor): *Pre-Cambrian and Lower Palaeozoic rocks of Wales*, 67–116. Cardiff.

BOSWELL, P. G. H. 1961. The case against a Lower Palaeozoic geosyncline. *Liv. & Manch. Geol. J.*, **2**, 612–24.

COOK, A. H. and THIRLAWAY, H. I. S. 1955. The geological results of measurements of gravity in the Welsh borders. *Quart. J. Geol. Soc.*, **111**, 47–70.

GEORGE, T. N. 1940. The structure of Gower. *Quart. J. Geol. Soc.*, **96**, 131–98.

—— 1962. Devonian and Carboniferous foundations of the Variscides in north-west Europe. *In* K. Coe (editor): *Some aspects of the Variscan fold belt*, 19–47. Manchester.

—— 1963. Palaeozoic growth of the British Caledonides. *In* F. H. Stewart and M. R. W. Johnson (editors): *The British Caledonides*, 1–33. Edinburgh.

GRIFFITHS, D. W. and GIBB, R. A. 1965. Bouguer gravity anomalies in Wales. *Geol. J.*, **4**, 335–42.

HANCOCK, P. L. 1964. The relations between folds and late-formed joints in south Pembrokeshire. *Geol. Mag.*, **101**, 174–84.

—— 1965. Axial-trace fractures and deformed concretionary rods in south Pembrokeshire. *Geol. Mag.*, **102**, 143–63.

JONES, O. T. 1912. The geological structure of central Wales and the adjoining regions. *Quart. J. Geol. Soc.*, **68**, 328–44.

—— 1924. Lead and zinc ores in the slaty rocks of Wales. *Trans. Inst. Min. Eng.*, **66**, 219–41.

—— 1956. The geological evolution of Wales and the adjacent regions. *Quart. J. Geol. Soc.*, **111**, 323–51.

OWEN, T. R. 1954. The structure of the Neath disturbance between Bryniau Gleision and Glynneath, South Wales. *Quart. J. Geol. Soc.*, **109**, 333–65.

POTTER, J. F. 1965. Rotational strike-slip faults, Llandeilo, Wales. *Geol. Mag.*, **102**, 496–500.

PRICE, N. J. 1962. The tectonics of the Aberystwyth Grits. *Geol. Mag.*, **99**, 542–57.

ROBERTS, J. C. 1966. A study of the reaction between jointing and structural evolution. *Geol. J.*, **5**, 157–72.

TROTTER, F. M. 1947. The structure of the Coal Measures in the Pontardawe-Ammanford area, South Wales. *Quart. J. Geol. Soc.*, **103**, 89–133.

11. Mesozoic Rocks

Intense denudation of the uplifted and corrugated land surface accompanied and followed the Hercynian movements. It continued into Triassic times, and before the region was again submerged thicknesses of the order of 10 000 to 15 000 ft of Palaeozoic rocks were removed from the anticlinal crests of the Vale of Glamorgan, Gower, and south Pembrokeshire. The consequent great break in the stratigraphical record, manifest in gross unconformity and in the absence of Permian and early Triassic (Bunter) deposits, is emphasized by the character of the flora and fauna of the overlying Mesozoic rocks, for during the long interval of erosion the Palaeozoic species died out, their places being taken by other kinds of organisms (*see* Fig. 36), of which ammonites and reptiles are particularly important.

The Mesozoic strata are limited in present outcrops to the ground south of the coalfield, and are well developed only in the Vale of Glamorgan. They comprise rocks belonging to two systems, the Trias represented by the Keuper and Rhaetic series, and the Jurassic by the Lias. The rocks are exceptional in southern Britain because of their deposition in close proximity to contemporary shores: they find their analogues elsewhere only on the flanks of the Mendip Hills. The first sediments are of Keuper age, South Wales having remained above depositional level in a zone of erosion during Permian and Bunter times. They were deposited in strongly transgressive overstep (*see* Pl. XA) on the subsiding floor of pared-down Palaeozoic rocks, in a great saline lake or gulf that extended eastwards and south-eastwards to cover a great part of southern England. With continued subsidence of its floor, the lake was ultimately invaded by the sea, transition sediments being represented by the Rhaetic Series, and fully neritic marine conditions being established in Liassic times.

With the subsidence, the lake waters and later the sea progressively drowned the land-mass of eroded Palaeozoic rocks to the north, stages of the drowning being very clearly shown by sustained overlap of younger beds across older. Details of the submergence are still well enough preserved to allow a reconstruction of the archipelago of islands about which the Mesozoic sediments accumulated, and of the retreating shorelines that are marked by the stages of overlap. (*See* Fig. 34.)

The rock-types are also signs of the depositional relations. As the rocks of each successive formation are traced towards the contemporary shore their lithology and their fossils change markedly, the contrasted facies they reveal being sharply distinctive of contrasted environments of sedimentation. Nearshore, the rocks are coarse-grained beach sands, gravels, and boulder beds, often massive and strongly lithified. Offshore, they are soft marls and shales, finer-grained and thinner-bedded, wide-ranging and showing little lateral variation in thickness. (*See* Fig. 35.)

Nearly all traces of the Mesozoic rocks have been removed from the ground to the west of the Vale of Glamorgan; but remnant evidence of the

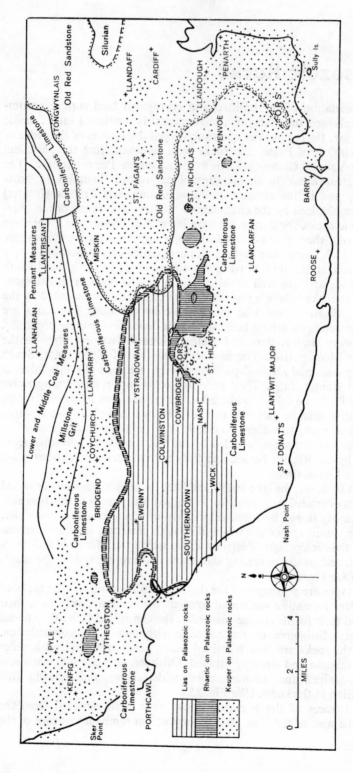

Fig. 34. *Map of the Mesozoic shorelines in the Vale of Glamorgan*

The archipelagic environment of sedimentation is indicated by the progressive drowning of the islands of Palaeozoic rocks as Mesozoic times advanced, the extensive upland to the west of Cowbridge not being completely submerged until Sinemurian times.

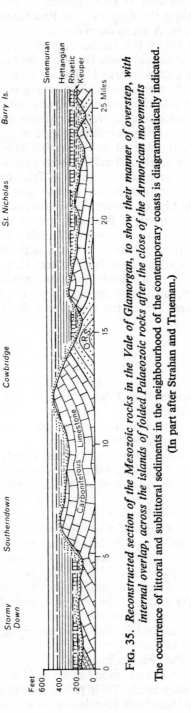

FIG. 35. *Reconstructed section of the Mesozoic rocks in the Vale of Glamorgan, to show their manner of overstep, with internal overlap, across the islands of folded Palaeozoic rocks after the close of the Armorican movements. The occurrence of littoral and sublittoral sediments in the neighbourhood of the contemporary coasts is diagrammatically indicated. (In part after Strahan and Trueman.)*

Triassic lake is to be seen as small pockets of red marls in some of the Gower bays, including the deep cave of Red Chamber excavated in Carboniferous Limestone. A more extensive outcrop of red breccias and conglomerates at Port Eynon is a further sign of a Triassic environment; and red-stained 'gash breccias' in south Pembrokeshire appear to be the tumbled walls of caverns in Carboniferous Limestone eroded in Triassic times. The repeated hints of a former cover of Trias along much of the present coast westwards from the Vale of Glamorgan suggest that the coastal profile as it is now seen is in part an exhumed profile originally carved by early-Mesozoic lakes and seas, much as the present landscape of the Vale itself is in part a Mesozoic landscape resurrected by the erosive late-Tertiary stripping of its Mesozoic cover after nearly 200 million years of burial.

Inland from the Vale of Glamorgan the Trias oversteps the Lower and Middle Coal Measures almost on to the Upper Coal Measures near Llantrisant, and the Lias in present distribution almost reaches the outcrop of the Coal Measures north of Bridgend; and although the early Mesozoic rocks appear in places to have impinged against a major shoreline in the barrier of a contemporary Pennant scarp, it is probable that the scarp was breached or submerged by later Jurassic seas that spilled widely over the coalfield. Direct signs of Mesozoic rocks are, however, unknown north of the present outcrops, unless they are to been seen in the red staining of some of the Coal Measures and in the veins and lodes of hematite in the Carboniferous Limestone.

Triassic Rocks

Keuper Series

In 'normal' facies the red Triassic rocks belong to the Keuper Marl, Bunter and Permian rocks being absent through overlap. They are about 500 ft thick in the Cardiff district, where they include 50 to 60 ft of conglomeratic basal beds and where they merge upwards into the Tea Green Marls, 25 to 30 ft thick, and the Sully Beds. Characteristically they consist of a uniform series of blocky mudstones with some silty layers. Many of the bedding planes are rain-pitted or sun-cracked, and carry other signs, including reptilian footprints, of deposition on a lake floor that was periodically subject to desiccation.

The red colour is a sign of a semi-desert environment, and the rocks generally are unfossiliferous—the lake waters were too saline, or seasonally too impermanent, to support much life. Salt pseudomorphs are found on some of the bedding planes, and although layers of rock-salt are not known to occur, nodules and beds of pink and white gypsum, particularly concentrated in the upper part of the sequence, indicate recurrent conditions of high evaporation in a playa environment. In all probability the lake was surrounded by an arid treeless hinterland, since there is a general absence of plants and insects as fossils.

The basal beds, the first sediments of the advancing lake waters, are littoral or nearshore rocks consisting mostly of breccias and conglomerates interbedded with sandy marls. Their pebbly constituents vary according to the lithology of the Palaeozoic foundation on which they rest. As the Red

Marls are followed westwards from the Cardiff district towards the island shores in the Triassic lake they pass laterally into littoral sediments like the basal beds: the coarser sediments are thus diachronous and may rise locally in the sequence, as in the outcrops west of Southerndown, to the uppermost levels of the Keuper series (*see* Fig. 35). Bluck has shown that the sedimentary structures and the grain-size distribution in some of the breccias and pebble-beds indicate fluctuating stream-flood and mudflow transport under highly variable climatic conditions typical of semi-desert environments; and along some of the steeper cliff slopes of the Triassic lake margin well-defined channels of transport may be recognized, the coarser poorly sorted gravels splaying out in fans at the channel mouths. The isolated pocket of breccias and conglomerates at Port Eynon is a spectacular example of such a fan poured into the hollow of an early-Mesozoic Port Eynon Bay.

The first signs of external influences on sedimentation in the Triassic playa lake are given by the Tea Green Marls, whose transition upwards from the hematite-stained Red Marls through an alternation of red and pale grey beds into a sequence of wholly grey beds points to a reducing environment of sedimentation and a progressive amelioration or replacement of the semi-desert salt-lake conditions of deposition. The Sully Beds, about 14 ft thick, are lithologically only the Tea Green Marls continued in upward sequence, distinguished because of the occurrence in them of the first fossil signs, the oyster *Liostrea bristovi*, of marine advance as the Keuper Series passes upwards into the Rhaetic.

Rhaetic Series[1]

The gradual lithological transition from lacustrine to marine sediments makes difficult a precise distinction of Rhaetic from Keuper: about Sully and St. Mary's Well Bay the topmost few feet of the Sully Beds, separated by a thin conglomerate from the similar but unfossiliferous rocks beneath, contain, in addition to *Liostrea bristovi*, the fishes *Hybodus minor* and *Gyrolepis alberti* of typical Rhaetic aspect, together with the labyrinthodonts *Mastodonsaurus* and *Trematosaurus*, and the bivalve *Rhaetavicula* [*Pteria*] *contorta* (*see* Fig. 36). The *bristovi* Beds are absent about Penarth and Lavernock, however, where there are a sharp lithological break and other signs of erosion and non-sequence at the top of the residual Sully Beds. The overlying Rhaetic rocks—sometimes called the Penarth Beds—are sharply distinct from any beneath: they fall into the two divisions of the Black or *Avicula contorta* Shales (the Westbury Beds) below, about 20 ft thick, and the White Lias above, about 11 ft thick.

The Black Shales are fine-grained thinly laminated brown, blue, and black shales with silty and gritty layers and a few thin ribs of impure limestones. They are underlain at Penarth by an inconstant gritty conglomerate, rarely more than an inch thick, that may equate with the conglomerate in the Sully Beds at St. Mary's Well Bay. It contains abundant fish scales, fish and reptilian bones and teeth, and coprolites—a characteristic Rhaetic 'bone bed'. Similar concentrations of vertebrate remains recur at several horizons in the Black Shales and indicate slow deposition. The dominant fossils of

---

[1]Though commonly regarded as the first stage of the marine Jurassic sequence, the Rhaetic (Rhaetian) is nowadays formally classed as Trias.

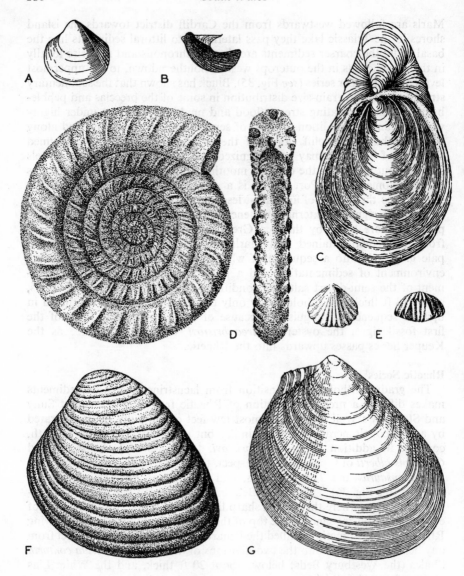

FIG. 36. *Fossils of the Mesozoic rocks*

(All natural size.)

A. *Protocardia rhaetica* (Merian), Rhaetic; B. *Rhaetavicula contorta* (Portlock), Rhaetic; C. *Gryphaea arcuata* Lamarck, Lower Lias; D. *Psiloceras* (*Caloceras*) cf. *bloomfieldense* Donovan, Lower Lias, two views; E. *Calcirhynchia calcaria* S. S. Buckman, Lower Lias; F. *Cardinia listeri* (J. Sowerby), Lower Lias; G. *Lima* (*Plagiostoma*) *gigantea* (J. Sowerby), Lower Lias.

B and F are from *British Mesozoic Fossils* published by the British Museum (Natural History). Permission to reproduce is acknowledged.

the Shales, however, are bivalves of many kinds, including *Rhaetavicula contorta*, *Chlamys valoniensis*, and *Protocardia rhaetica*. The fauna is restricted, nevertheless, for although some of the bedding planes are crowded with bivalves and ostracods, sometimes in an abundance to form thin shelly limestones, other groups of invertebrate fossils—corals, brachiopods, crinoids, ammonites, bryozoans—are very rare or lacking; and the general lithology of the Shales, richly pyritous and carbonaceous, suggests a relatively inhospitable environment perhaps of the poorly aerated waters of lagoonal flats.

The succeeding White Lias is mainly composed of calcareous marls and marly limestones showing signs in their light colour of refreshed waters of deposition and in their richer fauna of a less foetid habitat. Fossils are abundant in some of the beds: they include species of *Chlamys*, *Lima*, *Modiolus*, and *Eotrapezium* [*Schizodus*]. A bed towards the base, two or three feet thick, has been compared with the Cotham Beds of Somerset; and the overlying Langport and Watchet Beds may also be recognized.

Like the Keuper Marls beneath, the Rhaetic rocks of the Cardiff district (exposed in every detail in the cliffs between Penarth and Lavernock) were deposited in an environment relatively offshore. Westwards towards Cowbridge and Southerndown there is lateral transition to coarser pebbly beds and littoral conglomerates flanking the islands in the Mesozoic sea. In the neighbourhood of Bridgend, where the rocks transgress on to an Upper Carboniferous floor, massive gritty sandstones (the Quarella Sandstone) form much of the Rhaetic succession, their constituents derived in great part by the erosion and reworking of the sandstones in the Millstone Grit and Coal Measures. They fall into two main groups, the Lower Sandstone, about 20 ft thick, and the Upper Sandstone, about 35 ft thick, being separated by a mere 6 ft of residual Black Shales, and being overlain by Cotham Beds of the White Lias. They contain the gastropod *Natica pylensis* and abraded fish and reptile teeth and bones. As Francis has shown, their development is variable, the Lower Sandstone being locally transgressive across the Tea Green Marls and being itself overlapped near Coychurch by the Upper Sandstone. They show signs therefore not only of accumulation in close proximity to a nearby shore, but also of slight oscillations of sealevel and perhaps gentle warping as they accumulated.

Fissures, pipes and potholes in the Carboniferous Limestone of Glamorgan, were eroded during the earlier-Triassic evolution of a karst landscape; some of them, as marl-filled Neptunian dykes, descend several hundred feet below the old land surface. They sometimes contain trapped washings of Keuper or Rhaetic age that throw unexpected light on the contemporary land fauna. The washings may randomly be rich in the bones of small vertebrates, including small mouse-like symmetrodont and docodont forms that are mammal-like in their jaw structure and may be collateral ancestors of the pantotherian mammals of Jurassic times. One of these ancient links, *Morganucodon*, records its occurrence in its name.

Jurassic Rocks

The Jurassic system is represented in present outcrops, running from the Vale of Glamorgan into Gwent, only by the Lower Lias; but in no other part of Britain are the equivalent beds better displayed than in the spectacular cliffs and beach platform between Southerndown and Lavernock.

Over much of the ground, the Lias in 'normal' neritic facies (as 'Blue Lias') appears to follow the White Lias conformably, the junction lying where the cream-coloured tough uppermost marly limestone of the Rhaetic suite gives place to grey and brown shales (some of them 'paper shales' in their fine lamination) and thin impure limestones, some 45 ft thick, with *Psiloceras planorbis, Caloceras johnstoni*, and specimens of *Liostrea hisingeri* [*Ostrea liassica*] of early Hettangian age. These underlie about 100 ft of shales with some thin limestones (the lower part the Lavernock Shales) of later Hettangian age, in which ammonites are common. These in turn are overlain by alternations of fossiliferous nodular limestones and shales of Sinemurian age, in a thickness of about 200 ft. The rocks closely match their equivalents in Somerset and Dorset and are the western representatives of an extended basin of Jurassic sedimentation. No rocks younger than Sinemurian are known to occur, the *Psiloceras planorbis, Alsatites liasicus*, and *Schlotheimia angulata* zones forming the lowest 160 ft of the sequence, the *Arietites* [*Coroniceras*] *bucklandi* Zone forming almost the whole of the remainder (about 150 ft) and being capped by a few feet of the *Arnioceras semicostatum* Zone. In addition to ammonites characteristic of the zones, including *Charmasseiceras, Waehneroceras*, and *Metophioceras*, fossils are abundant and include belemnites, nautiloids, many bivalves (*Cardinia ovalis, Hippopodium ponderosum, Lima* (*Plagiostoma*) *gigantea, L.* (*P.*) *dunravenensis, Pseudolimea hettangiensis, Pinna hartmanni, Chlamys spp., Parallelodon hettangiensis, Pleuromya galathea, Pteromya tatei, Modiolus sp.*, and especially oysters, of which the incurved *Gryphaea* swarms in some of the beds), gastropods (*Pleurotomaria, Turritella*), the hexacorals *Montlivaltia* and *Stylophyllopsis*, rhynchonelloid and terebratuloid brachiopods (including *Calcirhynchia calcaria*), echinoid spines, and the crinoid *Pentacrinus*. Bones of ichthyosaurs and plesiosaurs are not uncommon. (*See* Fig. 36.)

As Hallam and Wobber have shown, the uniformity in superficial appearance presented by the alternations of shale and limestone in the Lias hides marked differences in detailed lithology between one bed and another. Richly fossiliferous layers full of shelly debris contrast with layers in which fossils are virtually absent. Some beds are riddled with 'worm'-borings and carry trails and tracks of a variety of invertebrate scavengers. The darker shales are bituminous and sometimes carbonaceous and suggest anaerobic conditions, the limestones suggest refreshed waters. Repeatedly there are signs of cyclothemic deposition that may reflect periodic climatic changes (perhaps seasonal in laminated shales) or epeirogenic changes in sea-level.

Trueman showed that the Hettangian rocks, traced from the Lavernock–Nash Point cliffs (*see* Pl. XB) westwards to Southerndown and northwards to Cowbridge, pass laterally into littoral deposits, diachronous in their development as upper beds overlap lower against a Liassic shelving shore. In this littoral facies two main rock-types occur, the Sutton Stone and the Southerndown Beds. The Sutton Stone is a massive light grey coarsely conglomeratic rock, mainly composed of detritus from the Carboniferous Limestone, very pure in its calcareous content. It is thick, a characteristic boulder-beach deposit, where it is banked against cliffs of Palaeozoic rocks, but it rapidly thins offshore to give place to the gravels of the Southerndown Beds. Locally it contains numbers of inshore kinds of fossils uncommon elsewhere—the massive corals *Isastraea* and *Thecosmilia*, thick-shelled coarsely ribbed scallops (*Chlamys suttonensis*), nests of brachiopods, large gastropods, bryozoans, sea-urchins, barnacles, and serpulids. The Southerndown Beds are relatively well-stratified and are interbedded with normal shales and limestones: they consist of pebbly and gritty lenticles, with rounded fragments (including rolled fossils and much chert) derived from the Carboniferous rocks, that wedge out offshore and form a laterally merging intermediate facies between the massive Sutton Stone and the Blue Lias. Their fossils include many Blue Lias forms, but gastropods, including the turreted *Katosira* and *Procerithium*, are unusually common.

The stages of overstep and overlap and the lateral changes of facies become less and less obvious in upward sequence. The coarser sediments are mainly of Hettangian age. By Sinemurian times, most of the islands in the Liassic sea had become submerged, and the deposits of the *bucklandi* Zone have an almost unbroken lithological uniformity in present outcrops. The absence of sandstones from the Lias where it lies within a mile or two of the Upper Carboniferous outcrops near Bridgend and Llantrisant has led Wobber to think that even the Pennant escarpment may for much of its length have been submerged by the Sinemurian (perhaps the Hettangian) seas, whose shores lay far inside the coalfield.

Glamorgan compares closely with Somerset and Dorset in the character of its Liassic rocks. The possibility that, as in Somerset and Dorset, younger Jurassic formations may at one time have extended into Wales and overlapped the Lower Lias is strengthened by the recent discovery that Middle and Upper Jurassic sediments, including Kimmeridge, occupy a syncline in the floor of the Bristol Channel only a few miles from the Vale of Glamorgan; and the proving in the Mochras borehole near Harlech of thick Upper and Middle Lias on Lower Lias, in a sequence that probably spreads widely over the floor of Cardigan Bay, has similar implications.

References

BLUCK, B. J. 1965. The sedimentary history of some Triassic conglomerates in the Vale of Glamorgan, South Wales. *Sedimentology*, **4**, 225–45.

CRAMPTON, C. B. 1960. Petrography of the Mesozoic succession in South Wales. *Geol. Mag.*, **97**, 54–65.

DONOVAN, D. T. and OTHERS. 1961. Geology of the floor of the Bristol Channel. *Nature*, **189**, 51–2.

FRANCIS, E. H. 1959. The Rhaetic of the Bridgend district, Glamorgan. *Proc. Geol. Assoc.*, **70**, 158–78.

HALLAM, A. 1960. A sedimentary and faunal study of the Blue Lias of Dorset and Glamorgan. *Phil. Trans. Roy. Soc.*, (B), **243**, 283–322.

LLOYD, A. J. 1963. Upper Jurassic rocks beneath the Bristol Channel. *Nature*, **198**, 375–6.

LUCY, W. C. 1886. Southerndown, Dunraven, and Bridgend beds. *Proc. Cottes. Nat. Field Club*, **8**, 254.

MISKIN, F. F. 1919. The Triassic rocks of south Glamorgan. *Trans. Cardiff Nat. Soc.*, 17–25.

MOORE, C. 1867. On abnormal conditions of Secondary deposits when connected with the Somersetshire and South Wales coal basin; and on the age of the Sutton and Southerndown series. *Quart. J. Geol. Soc.*, **23**, 449–568.

RICHARDSON, L. 1905. The Rhaetic and contiguous deposits of Glamorganshire. *Quart. J. Geol. Soc.*, **61**, 385–414.

THOMAS, T. M. 1952. Notes on the structures of some minor outlying occurrences of littoral Trias in the Vale of Glamorgan. *Geol. Mag.*, **89**, 153–62.

—— 1969. The Triassic rocks of the west-central section of the Vale of Glamorgan. *Proc. Geol. Assoc.*, **79**, 429–39.

TRUEMAN, A. E. 1920. The Liassic rocks of the Cardiff district. *Proc. Geol. Assoc.*, **31**, 93–107.

—— 1922. The Liassic rocks of Glamorgan. *Proc. Geol. Assoc.*, **33**, 245–84.

—— 1930. The Lower Lias (*bucklandi* Zone) of Nash Point, Glamorgan. *Proc. Geol. Assoc.*, **41**, 148–59.

WOBBER, F. J. 1966. A study of the depositional area of the Glamorgan Lias. *Proc. Geol. Assoc.*, **77**, 127–37.

—— 1968. Microsedimentary analysis of the Lias in South Wales. *Sed. Geol.*, **2**, 13–49.

—— 1968. Faunal analysis of the Lias (Lower Jurassic) of South Wales (Great Britain). *Palaeogeogr., Palaeoclimatol., Palaeoecol.* **5**, 269–308.

WOODWARD, H. B. 1893. The Jurassic rocks of Britain. Vol. iii: The Lias of England and Wales. *Mem. Geol. Surv.*

12. Tertiary Era

Whatever rocks of Mesozoic age younger than the Lias may have been deposited in South Wales, it is virtually certain they did not include Portland and Purbeck beds of the Upper Jurassic or Neocomian and Aptian beds of the Lower Cretaceous. Since Ramsay's day, over a hundred years ago, it has, however, been a recurrent hypothesis, emphasized notably by Jukes-Browne and Strahan, that the Chalk, deposited in a widespread transgressive sea, at one time extended over much of southern and central Wales, perhaps only the highest mountain summits—Plynlimon, the Brecon Beacons, Craig-y-Llyn—projecting above its surface as islands in the Upper Cretaceous sea. To this late-Mesozoic marine advance and erosive planing has been attributed the regularity of profile of the high plateau, interpreted as the relic of the Cretaceous sea-floor, from which all Cretaceous deposits have since been stripped by Tertiary erosion. The hypothesis is an attractive one, but it is without any direct stratigraphical or structural evidence in support, and there is some evidence, in the fragmentary Tertiary deposits that remain and in the folding of the Mesozoic rocks, that contradicts it.

Tertiary Deposits
The evidence of the rocks of southern England is conclusive that during the Eocene and Oligocene periods the sea retreated to the south and east of Wales; but it is probable, as the record of the Mochras borehole suggests, that Palaeogene limnic basins spread over parts of the Welsh mainland south-eastwards from Cardigan Bay. The succeeding Miocene was the period when the last of the intense orogenic movements affected Britain. The only deposit in South Wales that may possibly be ascribed to one of these periods is found in a small pocket of mottled and striped greasy plastic unfossiliferous clay ('pipeclay'), at least 45 ft thick, on the limestone plateau at Flimston in south Pembrokeshire. It has been compared by Dixon with the Bovey Tracey Beds of Devon, which also consist in part of plastic pottery clays, and like them it may have accumulated during Oligocene times as the fine sediment of a local freshwater lake. It rests directly on a floor of Carboniferous rocks, without sign of intervening Chalk, at a level far below the monadnock remnants of the higher plateau that form the summit peaks of Mynydd Preseli 1500 ft higher to the north; and it must owe its position either to deep erosion of the Mesozoic rocks in Eocene times or to down-warping in post-Oligocene times, in neither case in accordance with a postulated Cretaceous origin of the plateau.

Earth-movements
The Mesozoic rocks of the Vale of Glamorgan have been subjected to broad folding, with some faulting, ascribed by O. T. Jones to mid-Tertiary ('Miocene') deformation like that seen farther east in the 'Alpine' structures of the London and Hampshire basins. Although the deformation had an

inappreciable effect on the much more acute Armorican folds of the underlying Palaeozoic rocks, it was sufficient to throw the Triassic and Liassic
strata into anticlines and synclines of amplitude reaching several hundred
feet and to fracture them along faults several hundred feet in throw—
sufficient, that is, to have had an obvious effect on any erosional platforms
that may have existed at the time of the deformation. The finding of
Kimmeridge Beds at depth under the Bristol Channel to the south is a sign
of a complementary synclinal fold of still larger amplitude, post-Jurassic
in age. If it is supposed that Chalk even at maximum thickness once covered
the Kimmeridge Beds, its highest members would reach an altitude far below
the level of any part of the high plateau in southern Wales, and not much
above the level of some of the coastal platforms. The inferred extension of
thick Palaeogene sediments without underlying Chalk, from Mochras in a
basin occupying much of Cardigan Bay, downfaulted against the oldland
rocks of mid-Wales, is analogous evidence of the unlikelihood of the
plateau's being the exhumed floor of a Cretaceous sea.

Whether or not the Chalk covered much of South Wales at the close of
Cretaceous times, it is thus probable that the present major land-forms are
the product of Tertiary processes, in great part of post-Palaeogene processes.

Major Land-forms

In regional development the lower ground in South Wales, although it
has been much dissected by rivers, recognizably falls into a series of stepped
platforms rising from the present beach. The platforms bevel both the mild
structures in the Mesozoic rocks and the much sharper structures in the
Palaeozoic. The most impressive of them lies at a height of the order of
200 ft, although it may rise without notable break to about 300 ft, and fall
to about 160 ft. It is found in the Vale of Glamorgan planed in Liassic and
Carboniferous rocks, in Gower in Carboniferous rocks, and in Pembrokeshire
in both Upper and Lower Palaeozoic rocks (*see* Pls. IIA, IIIB, VB, VIB).
A higher platform at a little above 400 ft is seen in the Vale of Glamorgan,
in Pembrokeshire, and along the shores of Cardigan Bay, but it is not
easily recognized in Gower because of the encroachment of the lower platform. A third, at about 600 ft above sea-level, is well marked in northern
Pembrokeshire and in Cardiganshire, the crests of the dolerite monadnocks of
Carn Llidi and Pen Beri near St. David's being outlying relics (*see* Pls. IIA,
IIIB). In Gower the isolated summits in Old Red Sandstone—Cefn Bryn,
Rhossili Down, Llanmadoc Hill—also reach approximately the same height
and serve to confirm the continuity of the feature eastwards, despite great
degradation.

All these platforms, maintaining their essential form and relations (in
variant expression) along several hundred miles of coast, are characteristic
wave-cut benches uplifted by pulsatory emergence without significant
tilting or warping. Their disregard of the underlying geological structure
(Caledonian, Hercynian, or Alpine) makes them later-Tertiary (Neogene:
'Pliocene'—although it is possible the youngest of them may be Pleistocene)
in origin; and they correlate with similar platforms at about the same
heights that in southern England cut across folded Palaeogene rocks. The

quartz gravels scattered on the '200-foot' platform in south Pembrokeshire may be contemporary deposits, like the comparable 'Pliocene' gravels of the London basin.

These well-preserved coastal platforms offer analogues that help in interpreting degraded platforms at higher levels, finally in 'explaining' the high plateau. As Brown has shown, a stepped profile comprising three main elements may be recognized above the coastal platforms up to heights of about 2000 ft. The steps are often indistinct, with the steeper slopes between them much pared down, and with intermediate steps forming an irregularly multiple profile; but a discordance between surface and structure is almost universal. Brown has called the higher flats at each major step peneplains, and has regarded them as of subaerial origin; but their persistence over wide areas and their sustained uniformity of general height are as well or better explained by referring their origin, like the origin of the coastal platforms, to wave erosion and to pulsed uplift of a sequence of marine benches.

Clearly, this kind of upland profile, ranging through 2000 ft in height, cannot be the exhumed floor of a Cretaceous sea stripped of its cover of Chalk. In the absence of significant deformation it is presumably (although there is no direct evidence) later than the Miocene earth-movements and is no more than an upward continuation of the coastal profile: the combined elements form a unitary suite of which the higher are the older. The summits of Drygarn and Plynlimon in the moorlands of mid-Wales, and the escarpment crests of the Brecon Beacons and the Fans, which rise much above 2000 ft, are too remote and isolated to be integrated confidently into a synthetic 'plateau'; but in derivative and consonant hypothesis, they may be regarded as the last residuals of formerly extensive surfaces that have now been extinguished partly by the encroachment of lower benches, partly by river dissection.

River System

The emergent platforms, wave-cut benches eroded indifferently across a variety of rocks and structures, were the surfaces on which a primary river system was first established. With each pulse of uplift, and the exposure of lower benches, the youthful streams were extended seawards down the gentle platform slopes, in directions broadly normal to the successive shorelines. The pattern of drainage, long recognized by Lake and Strahan (who thought the surface of origin was an uplifted Chalk cover), may be seen in the elements of the valley courses that run fanwise from the core of the high plateau in mid-Wales—the Onny, the Teme, the Wye to Talgarth, the rivers of the Black Mountains, the Epynt rivers, the tributaries of the upper Towy, the coalfield rivers.

The primary river courses thus bear little relationship to the underlying geology, and are characteristically superimposed. They have become disrupted in later evolution by a secondary adjustment of tributaries to structure, in accelerated erosion along belts of weak or soft rocks. Piracy is a usual product of such evolution, and a number of the larger rivers in South Wales are subsequents. Thus the uppermost Rheidol–Teifi was diverted by rapid headward erosion of the upper Ystwyth along the line of the Ystwyth fault; and the uppermost Clywedog–Severn by deep-cutting headward erosion of the Twymyn–Dyfi along the line of the Twymyn fault—a capture seen in a specactular gorge and waterfall at the Dylife elbow. Similarly the lower Towy, the Tawe, and the Neath are all fed by right-bank tributaries that are matched by left-bank windgaps marking stream routes before capture: secondarily adapted to structure, the lower Towy flows along the axis of the Towy anticline, the Tawe and the Neath in faulted belts.

The pulsed emergence of the platforms caused repeated lowering of the base-level to which the rivers were graded. Many valleys, mature in their uppermost reaches, show rejuvenation downstream in the occurrence of nick-points, rapids and waterfalls (*see* Pl. XIIB), terraces, and steep-sided gorges. First illustrated in the Towy by O. T. Jones, residual profile-segments, extrapolated to inferred contemporary base-levels, repeatedly imply uplifts of about 580 ft, 420 ft, and 200 ft in rivers ranging from the Rheidol, Ystwyth, and Teifi in the west to the Towy, Tawe, and Neath in the south, and to the Usk, Wye and Upper Severn in the east. The implication is clear that over the whole area uplift was accompanied by insignificant tilting and that the river profiles may be correlated with the coastal platforms at corresponding heights. Rejuvenation stimulated piracy by the subsequent tributaries, and windgaps commonly grade with projected profiles of high-level river segments: thus windgaps show that the upper Towy flowed into the upper Usk at the time of emergence of the coastal platform at about 600 ft, and that the lower Towy and the Tawe as continuous rivers are very young and did not fully come into being until the coastal platform at about 200 ft was uplifted.

Repeated uplift and the trenching of valleys also influenced the distribution of ground water and the level of the water-table: Thomas's work makes clear how cave systems and solution subsidences in the Carboniferous Limestone reveal régimes of underground flow, now extinct, presumably of Neogene inception.

References

BROWN, E. H. 1952. Erosion surfaces in north Cardiganshire. *Trans. Inst. Brit. Geogr.*, No. 16, 51–66.
—— 1952. The River Ystwyth, Cardiganshire: a geomorphological study. *Proc. Geol. Assoc.*, **63**, 244–69.
—— 1956. The 600-foot platform in Wales. *Proc. 8th gen. Ass. 17th Int. Geogr. Union*, 304–12.
—— 1957. The physique of Wales. *Geogr. J.*, **126**, 318–34.
—— 1960. *The relief and drainage of Wales*. Cardiff.
DRISCOLL, E. M. 1958. The denudation chronology of the Vale of Glamorgan. *Trans. Inst. Brit. Geogr.*, No. 25, 45–57.
GEORGE, T. N. 1938. Shoreline evolution in the Swansea district. *Proc. Swansea Sci. & Field Nat. Soc.*, **2**, 23–48.
—— 1942. The development of the Towy and upper Usk drainage pattern. *Quart. J. Geol. Soc.*, **98**, 89–137.
—— 1961. The Welsh landscape. *Sci. Progr.*, **49**, 242–64.

GOSKAR, K. L. 1935. The form of the High Plateau in South Wales. *Proc. Swansea Sci. & Field Nat. Soc.*, **1**, 305–12.

—— and TRUEMAN, A. E. 1934. The coastal plateaux of South Wales. *Geol. Mag.*, **71**, 468–79.

HOLLINGWORTH, S. E. 1938. The recognition and correlation of high-level erosion surfaces in Britain. *Quart. J. Geol. Soc.*, **94**, 55–84.

JONES, M. 1949. The development of the Teifi drainage system. *Geography*, **34**, 136–45.

JONES, O. T. 1924. The Upper Towy drainage system. *Quart. J. Geol. Soc.*, **80**, 568–609.

—— 1930. Some episodes in the geological history of the Bristol Channel region. *Rep. Brit. Assoc.*, 57–82.

—— 1952. The drainage system of Wales and the adjacent regions. *Quart. J. Geol. Soc.*, **107**, 201–25.

JONES, R. O. 1931. The development of the Tawe drainage. *Proc. Geol. Assoc.*, **44**, 305–21.

—— 1939. The evolution of the Neath–Tawe drainage. *Proc. Geol. Assoc.*, **50**, 530–66.

LAKE, P. 1934. The rivers of Wales and their connection with the Thames. *Sci. Progr.*, no. 113, 23–40.

MILLER, A. A. 1937. The 600-foot platform in Carmarthenshire and Pembrokeshire. *Geogr. J.*, **90**, 148–59.

RICE, B. J. 1957. The erosional history of the Upper Wye basin. *Geogr. J.*, **123**, 357–70.

—— 1957. The drainage pattern and upland surfaces of south-central Wales. *Scot. Geogr. Mag.*, **73**, 111–22.

SAVIGEAR, R. A. G. 1953. Some observations on slope development in South Wales. *Trans. Inst. Brit. Geogr.*, no. 18, 31–51.

STRAHAN, A. 1902. On the origin of the river system of South Wales, and its connection with that of the Severn and the Thames. *Quart. J. Geol. Soc.*, **58**, 207–25.

THOMAS, T. M. 1959. The geomorphology of Brecknock. *Brycheiniog*, **3**, 55–136.

—— 1959. Solution subsidence outliers of Millstone Grit on the Carboniferous Limestone of the north crop of the South Wales coalfield. *Geol. Mag.*, **91**, 220–46.

—— 1963. Solution subsidence in south-east Carmarthenshire and south-west Breconshire. *Trans. Inst. Brit. Geogr.*, no. 33, 45–60.

WOOD, A. 1959. The erosional history of the cliffs around Aberystwyth. *Liv. & Manch. Geol. J.*, **2**, 271–87.

WOOLDRIDGE, S. W. 1950. The upland plains of Britain, their origins and geographical significance. *Adv. Sci.*, **7**, 162–75.

13. Pleistocene and Recent Deposits

Glaciation

The only widespread deposits in South Wales younger than the Lias are the tracts of unconsolidated glacial drift that mask the 'solid' rocks over much of the upland and the greater part of the lowland areas. These superficial accumulations of boulder clay and morainic gravel record a Pleistocene period of arctic climate when South Wales was covered by a sheet of ice that finally melted perhaps no more than 10 000 or 12 000 years ago.

As climatic conditions deteriorated with the onset of glaciation, snow compacting to ice accumulated on the higher hills as local ice-caps that, increasing in size, fed glacier tongues flowing into lower ground and coalescing as wide spreads of piedmont ice. At maximum glaciation the valley flanks became submerged, ice-sheds were less and less clearly defined, and expanding local ice-caps fused, to bury the whole country beneath an ice-cover that in places was several thousand feet thick.

Glacier movement conformed broadly with late-Tertiary land-forms, the chief outlets being along the major valleys (*see* Fig. 37). From the Plynlimon–Drygarn range much ice moved directly into the depression of Cardigan Bay, its main distributaries being the Dyfi, Rheidol, Ystwyth, Aeron, and Teifi glaciers; but there was also a considerable landward flow, on the one hand into the Severn valley to the north-east, on the other into the low ground about Rhayader and Builth to the south-east, from which the ice escaped mainly along the Wye valley into the plain of Hereford; and some ice, following the Irfon valley by Llanwrtyd, at times spilled south-westwards over the divide to feed the large Towy glacier that debouched into Carmarthen Bay.

Mynydd Epynt carried a small ice-cap that contributed eastwards and south-eastwards to the Wye and Usk glaciers, and north-westwards to Irfon ice that escaped into the Towy valley. Repeatedly, however, Plynlimon–Drygarn ice was powerful enough to override the Epynt scarp and to flow down the Epynt valleys into the Usk. Some Radnor ice as an Ithon tongue also joined the Wye glacier, but dominant movement from Radnor Forest was directly into the plain of Hereford. Although the details of fluctuating ice-pressures and ice-flows are not yet fully known, it is clear that mid-Wales was a region of convergent glaciers where there were not uncommon diversions, even reversals, of flow as temporarily a larger ice-mass impinged on a smaller; and some anomalous cols that cut through present-day watersheds may be a product of gouging by glacial overspill from ice-congested valleys lacking a free outlet.

Farther south a principal collecting ground was the escarpment of the Carmarthenshire Fans and the Brecon Beacons, whose steep northern face was the source of numerous corrie glaciers (*see* front cover and Pl. XIIA). Some ice flowed into the Usk and merged with ice from Mynydd Epynt and from the Black Mountains, some flowed north-westwards to

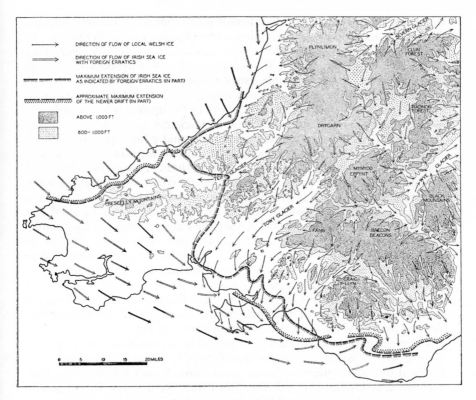

FIG. 37. *Map illustrating the glaciation of South Wales*

Arrows are generalized and no discrimination is made between the slightly
differing directions of ice-flow of Older-Drift and Newer-Drift times. It is not
certain that at its maximum extension the Irish-Sea ice covered or nearly
covered Mynydd Preseli, and its limit between the Teifi valley and the south
coast is conjectural.

(In part after Charlesworth, Williams, Dwerryhouse and Miller, and Griffiths.)

join the Towy glacier. The long dip-slopes towards the south were the
feeding-ground of large masses of ice, which, as powerful glaciers and sheets,
were able to cross much of the upland of the coalfield. The only barrier that
effectively shouldered off this Fans–Beacons ice-flow, and nourished a small
ice-cap of its own, was the Pennant escarpment dominated by Craig-y-Llyn
(*see* Pl. VIIB), which itself poured ice into the Neath and Cynon and on its
back slopes into the Afan, Llynfi, Ogmore, and Rhondda valleys.

The major outlets for the southward-flowing composite ice-drift were the
Loughor, Tawe, and Neath valleys in the west, and the Taff, Rhymney, and
Ebbw valleys in the east: the 'local' ice of the Gwendraeth and Loughor
valleys was also joined by overspills from the Towy glacier, particularly
through the low col near Llandebie into the Loughor valley, where erratics
of Llangadock rhyolite are common. Piedmont ice formed by the fusion of
these separate flows covered the coastal lowlands of Carmarthenshire, Gower,

and the Vale of Glamorgan. It extended for an unknown distance into the Bristol Channel and may have impinged upon the coast of north Devon.

All the native Welsh ice-flows were strictly 'local' in character: they originated in the hills of South Wales and moved on to neighbouring lowland down the most easily accessible valley routes. This is abundantly proved by the erratics in the drifts, the great majority of which were derived from sources immediately identifiable in outcrops of the upstream hinterland.

But, while the local ice-cover grew in thickness and extent, great ice-flows from diverse sources were commingling and piling up in the depression of the Irish Sea and St. George's Channel to form one of the largest and thickest ice-fields affecting the British Isles. Flows from the field debouched southward, crossed lowland North Wales, and impinged upon the coast of Cardigan Bay and north Pembrokeshire. Under great pressure they were sufficiently powerful to rise to heights of several hundred feet along the coast of Cardiganshire, despite the outward thrust of Plynlimon–Drygarn ice. Farther south, where the influence of local ice was not so great, they were able to ascend the Teifi valley for a score of miles and to overflow the col between Llandyssul and Carmarthen to join the Towy glacier. In north Pembrokeshire they rose over the flanks of Mynydd Preseli and, spreading on to the low ground of south Pembrokeshire and along the depression of the Bristol Channel, moved eastwards across Carmarthen Bay to Gower and the Vale of Glamorgan. (*See* Fig. 37.)

The range of this far-travelled ice is proved by the occurrence in the glacial drifts of erratic boulders of rock-types, foreign to South Wales, that can be matched by rock-in-place in distant parts of the Irish-Sea hinterland. Thus a common type of drift pebble found in Pembrokeshire and Gower has its counterpart in the granites and other igneous intrusions of southern Scotland. Erratics of the Goat Fell granite of the Isle of Arran and of the characteristic Ailsa Craig riebeckite microgranite of the Firth of Clyde have been found on the Cardiganshire coast and in Gower. Rocks from the Lake District are rare in South Wales, but rhyolites and rhyolitic ashes and tuffs from North Wales are widely scattered. On the Carmarthenshire lowlands and in Gower erratics from Pembrokeshire are relatively abundant, characteristic pebbles of the quartz-feldspar porphyry of Ramsey Island being particularly common. The coastal drifts of Cardigan Bay and Gower have also yielded fragments of Mesozoic rocks and fossils (of Liassic and Middle Jurassic age) that could have come only from outcrops in northern Ireland, or in the Hebrides, or from a submerged synclinal basin in the Irish Sea, or perhaps from the floor of Cardigan Bay itself. Flints of Cretaceous (Chalk) age, probably from the same sources, are also scattered widely through the lowland drifts of South Wales, and are particularly common in the Vale of Glamorgan where far-travelled igneous erratics are comparatively rare (but where, in the neighbourhood of Bridgend and Cowbridge, pebbles from sources in Scotland, North Wales, and Pembrokeshire have been found). The most easterly occurrences of Irish-Sea drift at present known are around Cardiff, where flints and felsite erratics from North Wales occur.

Associated with the igneous erratics in the gravels of the Cardiganshire and north Pembrokeshire coasts are pockets of sand with marine shells of arctic type including *Macoma balthica*, *Arctica islandica*, and *Astarte semi-*

(A.1086)

A. Raised Beach near Brandy Cove, Gower

B. Gorge of the River Rheidol at Devil's Bridge

(A.389)

A. Old Red Sandstone escarpment at Llyn-y-Fan

(*Cambridge Uni*

B. Lower Clun-gwyn Waterfall on the River Mellte

(A.490

sulcata, that were dredged from the sea-floor and carried inland by the Irish-Sea ice. Similar shelly drift has also been found about Pencoed in the Ewenny valley near Bridgend.

Difficulties in understanding the mode of glaciation that arise from the occurrence of two major ice-streams in South Wales (Welsh ice and Irish-Sea ice) are increased when it is recognized that the glaciation itself was multiple: there were at least two main glacial phases, separated by a relatively genial interglacial phase when the country or much of it was ice-free. The most complete evidence of the composite nature of the glaciation occurs along the shores of Cardigan Bay and north Pembrokeshire, where Jehu and Williams showed that a lower and an upper boulder clay, both of glacial origin, are separated by an interbedded series of fluviatile sands and gravels deposited when there was no neighbouring ice cover. Charlesworth (his conclusions, however, partly challenged by Mitchell and others) has traced the relics—terminal moraines and associated outwash sands and gravels—of the second glacial phase (the Newer Drift) and has demonstrated that they do not cover such an extensive area as those of the first (the Older Drift): local Welsh ice reached the Bristol Channel only in Swansea and Carmarthen Bays; and, while Irish-Sea ice impinged upon the shores of Cardigan Bay and north Pembrokeshire, it failed to cross into Carmarthenshire and Glamorgan, where the Newer Drift consists only of locally derived material. Inland, where the parent sources and the routes of transport of the Newer Drift were the same as those of the Older, it is often difficult to distinguish between the deposits and the geomorphic effects of the two phases of glaciation; but, where the evidence is adequate, it is clear that the second glaciation was much less powerful than the first, Newer Drift of Fans–Beacons origin being mainly confined to the larger valleys and overriding only the flanks of the higher hills of the coalfield. Minor halt stages or temporary fluctuations in the final retreat of the ice-front are marked by small transverse moraines in many valleys.

The erosional effects of glaciation are strongly delineated in the modification of the pre-Glacial topography. The scarp faces of the Old Red Sandstone and the Pennant Sandstone are pocked by well-developed corries (*see* Pls. VIIB, XIIA); and corries form the heads of many of the valleys in the hills of mid-Wales. A number of them are occupied by moraine-dammed lakes—for instance, Llyn-y-Fan at the head of the Tawe, Llyn Fawr beneath the Pennant crest of Craig-y-Llyn, Llyn Llygad Rheidol on the north slope of Plynlimon. Gouging and over-deepening are seen in the U-shaped cross-profile of many valleys but few true rock basins are known.

Many of the pre-Glacial valleys are choked by drift, sometimes to a thickness causing them to be abandoned by post-Glacial rivers. Signs of temporary diversions, as dry meanders and spillways (*see* Pl. VIA), are common, and permanent diversions may be intricate and have a complex history, as O. T. Jones described in the Teifi valley. Dry courses of former streams or streamlets, sometimes without clear relationship to the regional drainage pattern and without obvious source or outlet, are found in many places and have been interpreted notably by Bowen as subglacial meltwater channels formed during the later stages of ice wastage and retreat and abandoned when the ice finally disappeared.

During the later stages of glaciation along the coast of Cardigan Bay, when the melting front of the local Welsh ice had withdrawn some distance inland, Irish-Sea ice still formed a slowly receding barrier across the river mouths. The free drainage of the area was consequently impeded, and the impounded meltwaters collected as temporary lakes which discharged southwards through overflow channels into neighbouring valleys (*see* Pl. VIA). Such lakes have been shown by Jones and Pugh to be well represented in the Aberystwyth district, but the largest, described by Charlesworth and by Jones, lay in the Teifi valley ('Lake Teifi') (*see* Fig. 38): it drained southwards into a lake at a lower level in the Nevern valley, which in turn overflowed at times through the present Gwaun valley into

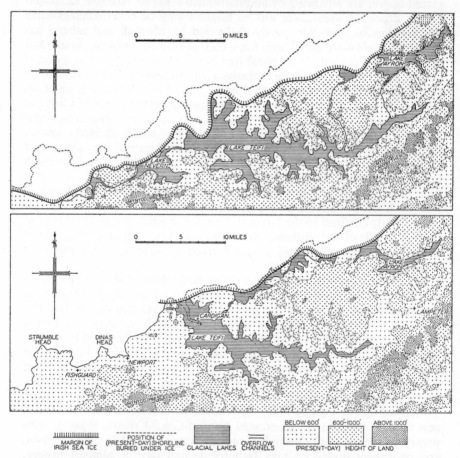

FIG. 38. *Late-Glacial ice-dammed lakes in Cardiganshire and north Pembrokeshire*

The upper map shows a stage when the Irish-Sea ice still lay relatively far inland, the ultimate outlet of the lake being then south-westwards *via* the Jordanston Channel into the upper reaches of the Cleddau. The lower map shows a later stage when the Irish-Sea ice had retreated locally beyond the coast, and the lakes were reduced in size and drained directly into Cardigan Bay.

(After Charlesworth and Jones.)

Fishguard Bay. It is probable that at the time of its initiation the drainage of this lake system to the sea was through a channel at Jordanston into the upper reaches of the Western Cleddau and so to Milford Haven.

Comparable periglacial lakes, dammed by Irish-Sea ice occupying the Bristol Channel, may have existed during the later stages of the Older-Drift glaciation. The largest of them is inferred to have occupied much of the low hinterland behind Carmarthen and Swansea Bays as 'Lake Taf', 'Lake Loughor', and 'Lake Tawe', at one stage forming a single confluent lake extending from Whitland to Bridgend: in Griffiths's reconstruction the lake limits are defined by the distribution of fine-grained erratic detritus of Irish-Sea origin, transported by lake currents. In alternative explanation, however, Bowen has postulated that the Irish-Sea detritus was dispersed not in a lake but by meltwaters flowing off Irish-Sea ice on to residual Welsh ice lying at lower altitudes to the north; and, in an absence of any clear signs of banded lake deposits and lake-shore beaches, he has suggested that what appear to be overflow channels of lake drainage (like the Dunvant and Cockett gaps near Swansea) were eroded by subglacial meltwater streams and follow accidental courses.

Raised Beaches

Reaching heights of 20 to 30 ft above present-day high-water mark, raised beaches and beach platforms occur intermittently wherever the nature of the coast is suitable for their formation and preservation. They are well displayed along the foot of the limestone cliffs forming much of the coast of Gower and south Pembrokeshire. They may also be recognized in the cliffs of St. Bride's Bay and Whitesand Bay, and of Grassholm in the far west. The most prominent of them, first recognized by Tiddeman, has been called the *Patella* Beach because of the abundance of limpets it contains. In Gower it is widespread as a deposit of shelly shingle (in which periwinkles and dog-whelks are also very common), generally cemented by stalagmitic calcite into hard conglomerate, and resting on a well-planed, though narrow, wave-cut platform (*see* Pl. XIA). More usually the platform may be recognized though no contemporary deposits remain. The beach in Gower is seen to be overlain by glacial gravels of the Older Drift. Nevertheless, pebbles in the beach include many erratics, some of them of igneous rocks and flint of distant origin: they may have been carried by drifting ice at a time of onset of arctic conditions. Similar erratics are found in the raised beach in Pembrokeshire.

Overlying the beach at a number of localities there is a mixed sequence of sediments, usually of no great thickness, including head, solifluxion and colluvial deposits, and wind-blown sands, in a variable and sometimes bewildering order. Except that they all (including the beach) are probably later-Pleistocene in age, there is as yet no detailed agreement on the stratigraphical relations of the deposits or on their correlations with English and Continental glacial phases. The beach itself is sometimes regarded as falling into the Hoxnian interglacial interlude, the Older Drift as representing a Welsh facies of either the main Würm or the earlier Riss (Saale) glaciation, and the Newer Drift the late-Würm (Weichsel) glaciation; but the direct evidence is slender.

Many of the bone caves of Gower and Pembrokeshire open on to the platform of the *Patella* Beach: it is probable that they were excavated, or at least enlarged, by marine action along faults and joints at the same time as the platform was planed. At that time the modern beach platform had not been formed, and prehistoric animals found easy access to the caves over the pavement of the *Patella* Beach.

In Gower (Minchin Hole) there is evidence of another beach, the *Neritoides* Beach, also probably older than the Older Drift, but overlying (and therefore younger than) the *Patella* Beach from which it is separated in places by a bone-bearing terrestrial cave-breccia.

A third ancient beach, the Heatherslade Beach in Gower, contains an abundance of erratic pebbles derived from the neighbouring glacial gravels, and it is certainly much younger than the Older Drift. Remnants of the beach, consisting of a concreted shelly conglomerate and shingle, are found below present-day high-water mark resting directly on the 'modern' beach platform. The platform therefore is not as recent in origin as it appears to be: it was planed no later than the time of formation of the beach, and has since undergone elevation and subsidence, finally to return coincidentally to its original level.

Bone Caves

Many of the caves of Gower, eroded in the Carboniferous Limestone and opening on to the platform of the *Patella* Beach, were inhabited by pre-historic animals at a time when the land was appreciably higher above sea-level than it is at the present day. The caves contain thin deposits of cave earth and breccia often rich in mammalian bones of Pleistocene age. The most important is Paviland, described by Sollas, who showed that it was occupied by cromagnard men (including the 'Red Lady') during Aurignacian, and possibly at the beginning of Solutrean, times—that is, during the temperate interval between the glacial phases of the Older and Newer Drifts. Artefacts associated with the few human bones (which appear to have been interred) include numbers of chipped flint flakes and some ornaments of ivory and bone, all characteristically later Palaeolithic. With the human bones were found bones of other animals including many of horse, bear, and ox, together with mammoth (*Mammuthus* [*Elephas*] *primigenius*), woolly rhinoceros (*Coelodonta antiquitatis*), Irish elk (*Megaceros hibernicus*), wolf, and reindeer. The fauna as a whole is of a cold-temperate type, suggesting that steppe conditions characterized the contemporary environment in South Wales.

Other caves in Gower contain many Aurignacian implements, but none has yielded human bones of Palaeolithic age. The implements are usually associated with a 'cold' fauna comparable with that of Paviland. In Bacon Hole and Minchin Hole (where it is associated with the *Neritoides* Beach), an older fauna without human relics includes the straight-tusked elephant (*Archidiskodon meridionalis*) and the slender-nosed rhinoceros (*Dicerorhinus hemitoechus*) that appear to have lived at a time when the climate was appreciably warmer than it is at present. A few of the caves, containing many hundreds of bones, were probably hyaena dens, and one, where over a thousand shed antlers were found, was a retreat of red deer.

Similar caves occur along the Pembrokeshire coast: the most important Palaeolithic site, Hoyle's Mouth near Tenby, has yielded implements of Aurignacian type associated with relics of a 'cold' fauna of reindeer, red deer, wolf, and bear. Abundant bones and teeth from caves and fissures in Caldey include remains of hyaena, cave lion, wolf, hippopotamus, mammoth, woolly rhinoceros, Irish elk, reindeer, and red deer.

Submerged Peats

The principal post-glacial deposits, excepting river alluvium, are found along the coast, where extensive tracts of blown sand, piled by the prevailing winds, occur particularly in Carmarthen and Swansea bays. The dunes impede drainage, and behind them there are stretches of flat alluvial marshland running parallel with the coast. The relationships are well displayed between Swansea and Porthcawl, east of Pendine, and on a smaller scale south of Kidwelly and at Oxwich. Farther north, Borth Bog at the mouth of the River Dyfi is protected from the sea by a sand spit several miles long.

For much of post-Glacial times similar conditions appear to have been widespread around the South Wales coast, many of the bays displaying evidence of ancient marsh conditions in the presence of lacustrine, estuarine, and terrestrial sediments. These strata, formerly known as the Submerged Forest Series, are usually exposed at or below high-water mark in positions where they could not possibly be formed at the present time. The terrestrial sediments are richly humic soils that have been compacted to form peat; stools of trees, still in position of growth, are common in them, and their mass is composed of branches, twigs, and leaves of such plants as oak, hazel, alder, and birch that must have grown on relatively dry ground at least a few feet above sea- and marsh-level.

Seen in Cardigan Bay (the traditional site of Cantref-y-Gwaelod) and in Pembrokeshire, peat beds have also been particularly well exposed in dock and other excavations in the Glamorgan ports, where they are interbedded with silts, muds, clays, and gravels. The details are illustrated in Fig. 39, from which it is clear that coastal South Wales suffered drowning in post-Glacial times. On the assumption that the lowest peat beds accumulated at a height of at least 20 ft above contemporary mean sea-level (so that the living plants were free from inundation by the sea), relative subsidence must have been at least 75 ft. The subsidence, a complement of the world-wide rise in sea-level—the Flandrian transgression of north-western Europe—that followed the melting of the ice at the close of Pleistocene times, was however not sustained and regular, but intermittent and pulsatory, with alternations of freshwater and even marine sediments interbedded with the peats: the freshwater strata contain not only reeds and sedges, but also such typical marsh snails as *Limnaea, Planorbis, Hydrobia,* and *Ancylus;* and the brackish-water estuarine and marine beds contain the bivalves *Scrobicularia* and *Macoma.* The present-day alluvium of the coastal flats is only the latest of these deposits, and the rise in sea-level, begun in early post-Glacial times, may not yet have come to an end, although no appreciable change in the position of sea-level has occurred during recorded history.

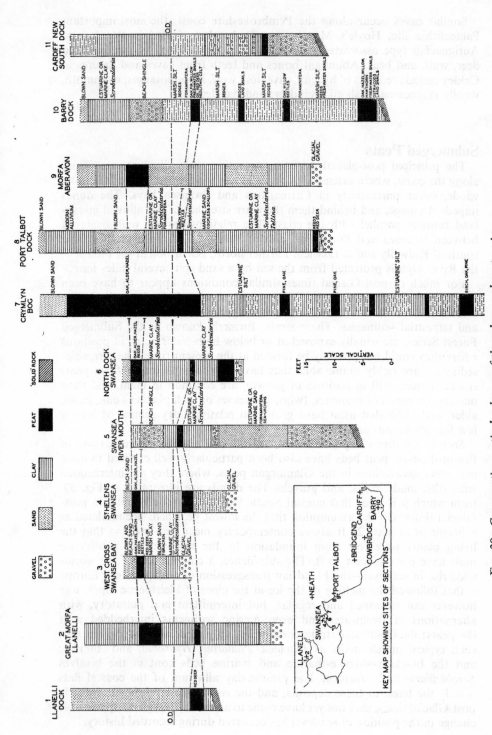

FIG. 39. *Comparative vertical columns of the submerged peat series*

In the thick sequence of the deposits proved in Crymlyn Bog, the peats record progression from 'Boreal' through 'Atlantic' to 'Sub-Boreal' climatic phases, as judged by the associations of plants (pollen) they contain. Described by Godwin, they span an interval of perhaps 8000 years, the youngest peat, over 20 ft thick with its surface now lying nearly 20 ft above mean sea-level, having begun to be formed perhaps 3000 years ago.

Associated with the plants in the peat beds are remains of insect life (the elytra of beetles are especially common) together with bones of mammals some of which—red deer, roe deer, auroch—no longer live wild in Wales. Flint implements and flint-working sites associated with the peat bed exposed in the Pembrokeshire and Glamorgan bays show that the forest land was then sufficiently dry and protected to be, at least temporarily, the habitation of man: the artefacts are typically Neolithic.

The drowning of most of the larger valleys around the coast of South Wales—the ria of Milford Haven is an outstanding example—is to be attributed in part to glacial over-deepening, in part to the Flandrian post-Glacial rise in sea-level (though some regional subsidence had already occurred before the close of Tertiary times). Locally the effect of submergence is obscured by the thick accumulations of drift and alluvium that choke the valley mouths; but boreholes and excavations, and geophysical probing, have proved the 'solid' rock floor to lie in places more than 150 ft below present-day sea-level.

Inland, the only notable Recent deposits are the river alluvium of most of the valleys and upland peats. The Teifi near Tregaron flows through one of the largest of the peat bogs, formerly the floor of a moraine-dammed lake, whose earliest post-Glacial deposits to which an age can be given are of 'Pre-Boreal' age, older (at about 10 000 years) than the oldest peat recognized at Crymlyn, and implying a final melting of the ice in mid-Wales not much earlier.

References

ALLEN, A. 1960. Seismic refraction investigations of the pre-Glacial valley of the River Teifi near Cardigan. *Geol. Mag.*, **92**, 276–82.

ANDERSON, J. G. C. and BLUNDELL, C. R. K. 1965. The sub-drift rock surface and buried valleys of the Cardiff district. *Proc. Geol. Assoc.*, **76**, 367–77.

BOWEN, D. Q. 1966. Dating Pleistocene events in south-west Wales. *Nature*, **211**, 475–6.

—— 1967. On the supposed ice-dammed lakes of South Wales. *Trans. Cardiff Nat. Soc.*, **93**, 4–17.

—— and GREGORY, K. J. 1965. A glacial drainage system near Fishguard, Pembrokeshire. *Proc. Geol. Assoc.*, **76**, 275–82.

CHARLESWORTH, J. K. 1929. The South Wales end-moraine. *Quart. J. Geol. Soc.*, **85**, 335–58.

CODRINGTON, T. 1898. On some submerged rock valleys in South Wales, Devon, and Cornwall. *Quart. J. Geol. Soc.*, **54**, 251–78.

CRAMPTON, C. B. 1966. Certain effects of glacial events in the Vale of Glamorgan, South Wales. *J. Glaciol.*, **6**, 261–6.

DAVID, T. W. E. 1883. On the evidence of glacial action in south Brecknockshire and east Glamorganshire. *Quart. J. Geol. Soc.*, **39**, 39–54.

DWERRYHOUSE, A. R. and MILLER, A. A. 1930. The glaciation of Clun Forest, Radnor Forest, and some adjoining districts. *Quart. J. Geol. Soc.*, **86**, 92–129.

GEORGE, T. N. 1930. The submerged forest in Gower. *Proc. Swansea Sci. & Field Nat. Soc.*, **1**, 100–8.

—— 1932. The Quaternary beaches of Gower. *Proc. Geol. Assoc.*, **43**, 291–324.

—— 1933. The glacial deposits of Gower. *Geol. Mag.*, **70**, 208–32.

—— 1933. The coast of Gower. *Proc. Swansea Sci. & Field Nat. Soc.*, **1**, 192–206.

—— 1936. The geology of the Swansea main drainage excavations. *Proc. Swansea Sci. & Field Nat. Soc.*, **1**, 331–60.

GODWIN, H. 1940. A Boreal transgression in Swansea Bay. *New Phyt.*, **39**, 308–21.

—— and MITCHELL, G. F. 1938. Stratigraphy and development of two raised bogs near Tregaron, Cardiganshire. *New Phyt.*, **37**, 425–54.

—— and NEWTON, L. 1938. The submerged forest at Borth and Ynyslas, Cardiganshire. *New Phyt.*, **37**, 333–44.

GRIFFITHS, J. C. 1939. The mineralogy of the glacial deposits of the region between the Rivers Neath and Towy, South Wales. *Proc. Geol. Assoc.*, **50**, 433–62.

JEHU, T. J. 1904. The glacial deposits of northern Pembrokeshire. *Trans. Roy. Soc. Edin.*, **61**, 53–87.

JOHN, B. S. 1965. A possible Main Würm glaciation in west Pembrokeshire. *Nature*, **207**, 622–3.

JONES, O. T. 1942. The buried channel of the Tawe valley, near Ynystawe, Glamorganshire. *Quart. J. Geol. Soc.*, **98**, 61–88.

—— 1964. The Glacial and post-Glacial history of the lower Teifi valley. *Quart. J. Geol. Soc.*, **121**, 247–81.

LEACH, A. L. 1911. On the relation of the glacial drift to the raised beach near Porth Clais, St. David's. *Geol. Mag.*, **48**, 462–6.

—— 1918. Flint-working sites on the submerged land (submerged forest) bordering the Pembrokeshire coast. *Proc. Geol. Assoc.*, **29**, 46–7.

LEWIS, C. A. 1966. The Breconshire end-moraine. *Nature*, **212**, 1559–61.

MITCHELL, G. F. 1960. The Pleistocene history of the Irish Sea. *Adv. Sci.*, **17**, 313–25.

—— 1962. Summer field meeting in Wales and Ireland. *Proc. Geol. Assoc.*, **73**, 197–214.

MOGGERIDGE, M. 1856. On the section exposed in the excavation of the Swansea docks. *Quart. J. Geol. Soc.*, **12**, 169–71.

SOLLAS, W. J. 1883. The estuaries of the Severn and its tributaries. *Quart. J. Geol. Soc.*, **39**, 611–26.

—— 1913. Paviland cave: an Aurignacian station in Wales. *Trans. Roy. Anthrop. Inst.*, **43**, 339–70.

STRAHAN, A. 1896. On submerged land-surfaces at Barry, Glamorganshire. *Quart. J. Geol. Soc.*, **52**, 74–89.

STUART, A. 1924. The petrology of the dune sands of South Wales. *Proc. Geol. Assoc.*, **35**, 316–31.

WATSON, E. 1966. Periglacial structures in the Aberystwyth region of central Wales. *Proc. Geol. Assoc.*, **76**, 443–62.

WEST, R. G. 1963. Problems of the British Quaternary. *Proc. Geol. Assoc.*, **74**, 147–86.

WILLIAMS, G. J. 1968. The buried channel and superficial deposits of the lower Usk, and their correlation with similar features in the lower Severn. *Proc. Geol. Assoc.*, **79**, 325–48.